"GROWING UP POOR" IN A COUNTRY OF PLENTY

by

Samuel Erasmus Ayettey

Some of the links presented in the book might become outdated as their functionality is subjected to their owner's inclinations. In such a case, readers are advised to search for alternatives.

Contents

Dedication

I dedicate this book to my parents, college professors, and the community of my upbringing in Osu, Accra, Ghana. Education was on the lips of every parent to encourage their children to achieve where I was raised. Substance abuse was not allowed because our parents knew it was bad for our health. Teachers, parents, and elders in the community were our role models. There are many children raised in substance abuse families afflicted with a painful history of abuse and neglect. These children are at risk of abusing drugs and alcohol to escape their hurt, anger, frustration, and drop out of school—violence and criminal activities in our community result from self-hatred and a lack of love among the afflicted. A majority of people in prison have no job skills or meaningful trade or vocational training. Vocational and trade programs in public schools would support at-risk students to get jobs and reduce the crime rate. This book aims to educate teachers, parents, and mental health workers to identify antisocial traits among the youth for treatment.

Acknowledgement

I personally wish to thank Steve Brown and the 360 Ghostwriting team for taking their time to help me finish this book. I greatly appreciate the fact that the writing team at 360 Ghostwriting helped to arrange and contributed to the creation of the book cover ,graphic design, editing and teach me how things are done in the publishing world. I cannot thank the writing team enough as co -author for helping me to writer this book. This book wouldn't have been possible without the input of 360 Ghostwriting team that allowed me to develop insight related to ideas and constructive comments for the chapters of this book. I want to thank the writing team also for their leadership to get the job gone. I cannot thank the team enough for being my unrelating source of inspiration to let me know how things get done in the publishing world. I also thank each my role model supervisors from past to present, college, coworkers on the job and those who helped me to develop my thought on the job to write this book.

Preface

While physical health and ailments have rarely been ignored by societies worldwide, it's mental health that has largely been ignored in some parts of the world. In some places, mental health is outrightly rejected as non-existent. It's ironic that mental health impacts physical health more than anything else. Stress is reportedly the number one cause of heart disease, and that is a mental health concern. In our societies, we are exposed to so many dangers that can impact our mental health, and one of those is substance abuse. As I will explain in the first chapter, substance abuse is another name for drug addiction. I have written this book to educate and create awareness to the masses about this evil that has plagued our societies. This one evil has destroyed families, homes and left its victims on the streets without jobs and sometimes even a home to go to. They would then resort to crime to feed their addiction and have no one to guide them and get them back on track. Our civic duty is very much to help such people overcome their addiction and become productive members of our society.

It is truly my hope that this book provides the necessary education so we can save ourselves and our loved ones from this menace. The book's point is to advise, educate, and create awareness of problems the youth face regarding substance and alcohol abuse. I would hope that this book brings about the much-needed change in our communities regarding substance abuse. Our high school students are vulnerable to it and fall into a trap if not educated in its dangers. Whether it's drugs or alcohol, abuse is abuse, and it needs to be set right. This book is also for parents to help teach their children the dangers of alcoholism and drug addiction. This education needs to start from the very beginning as we are responsible for raising our children right.

Our public schools will have some students drop out each year because they would resort to substance abuse, and they would then need guidance from mental health clinics, teachers, and parents. Our public schools should have substance abuse awareness programs to educate our youth, and this book is expected to play a role in that end. I want those addicts to be rehabilitated and brought back into our communities to earn a meaningful living. We wouldn't want them to go down the same self-destructive route again. They would end up in poverty, homeless and jobless and hence resort to criminal activities to keep up with their substance abuse. We need to rescue these people from this dark path and bring them back into the light to live happy and fulfilling lives. No marriage can survive if one of the spouses indulges in substance abuse. It almost always results in the end of the marital union. The worst thing that could happen to individuals involved in substance abuse is suicidal thoughts because their dependency on drugs and alcohol is way too much for them to handle. They end up getting lonely, depressed, and antisocial because they think that these drugs and alcohol are their best friends and saviors. However, they do not realise how much harm they're causing themselves, their loved ones, and their local communities. I can't stress enough that we need to save these individuals before becoming a casualty of their self-destructive behaviours.

This book is very personal to me, and you will realise it when you read through the chapters. I want to invite you all to see that individuals out there need our help in literally coming back to life. Crime, violence, and drugs are evils of our society and are very much interrelated than we could ever imagine. I see myself as an advocate for change because I want to put an end to substance abuse and alcoholism and the best way to do it is to create awareness. This book is going to discuss substance abuse in detail and how badly it has impacted our communities. I have worked with individuals who are involved in substance abusers and alcoholics. For this very reason, I want to save our youth from these evils that have plagued our communities.

I have been on an enlightening journey as I have healed those who would be considered society's outcasts and assisting them with rehabilitation in the Inner City. I have worked with these individuals so they could get a second chance at life as they are hopeless. I have given them hope and an opportunity to earn a decent living. Therefore, it is my mission to get schools to start trade and vocational programmes for former substance abusers and recovered addicts. These would also include those who have spent time behind bars and have been reintroduced to society, very much like continuing education programmes in universities worldwide. I want this book to go a long way in achieving my goals and dreams of a substance/ drug-free community wherever we live. We all have a duty and responsibility to these individuals because these are one of us and have gone down the downtrodden path. These lost souls need our assistance in being reintegrated into society. I am very confident that if they receive the right level of support and guidance, they could very well become valuable members of our communities, lead beautiful lives, and have families of their own. As you start reading from the first chapter of this book to the very end, I urge you to take substance abuse and alcoholism seriously, as these evils of society need to be eradicated for good. This is for our children, their children, and so on. Thank you very much for choosing to read this book, and it is my utmost hope that you gain the much-needed awareness and start your very own awareness campaigns. Here's to change!

Chapter 1: What is Substance Abuse?

Many people in our communities don't understand what substance abuse actually entails. Before we get into the nitty-gritty of it, I will explain the concept behind substance abuse. The idea here is that anything can become abuse if it is being used or consumed too much. We all have heard the saying: Too much of anything is bad for you. Therefore, the same idea applies to substance abuse or, in other words, drug abuse. However, substance abuse isn't necessarily just limited to illegal and hard drugs but even painkillers, prescription medications, cigarette smoking, and even alcoholism. Illegal drugs are also known as designer drugs and include hallucinogens, cocaine, heroin, ecstasy, LSD, Methamphetamine (also known as crystal meth), and more. Some parts of the world have legalised marijuana, although it is also considered an illegal drug. We will touch upon marijuana use and abuse later on in this chapter. When anyone consumes too much of these drugs, they become dependent on them. Once the dependence increases, it becomes a massive problem as withdrawal symptoms can become very difficult for the abuser.

To keep up their addiction and habit, they would become violent and argue with friends, family, and loved ones. They would be removed from their places of employment and lured into the world of crime. They would then become what many would term as a menace to society because they would resort to anything to keep fueling their addictions. These people know that these drugs are harming them, but they will still consume them. Addiction is a disease, according to WebMD, and this isn't further from the truth. The worst thing that could happen to these individuals is that they could die from their addiction. They would inflict so much harm on their bodies by fueling their addictions that they would die from them. We all have heard of people dying of drug overdose, and it's a severe issue. If these individuals could have been helped, they would have become valuable members of society and lead fulfilling lives.

The Pleasure in Pain

I was always fascinated to learn why individuals get addicted to drugs and alcohol and abuse them, knowing that these are slowly killing them. When I figured it out, I got pretty surprised, and I called the reason by the term "the pleasure principle." The idea behind this principle, in my opinion, is there they seek pleasure from their pain. The "buzz" or "high" from these drugs and alcoholic drinks gives them so much pleasure that they want to keep consuming them. Once they reach a stage that they can't afford their addictions and are craving the high, they turn to violence and crime. This is because these drugs have such a controlling effect on their personalities like they're zombies. They just are out for blood because they want to get their kicks, and that turns them into society's outcasts. They'll resort to crime and violence and get thrown in jail, but what happens after they're incarcerated. If they haven't gotten over their addictions, their cravings will not leave them. Due to the pleasure principle, the use becomes abuse which becomes a massive problem for everyone in their communities.

The pleasure principle can be applied to anything, but the effects aren't ad adverse as drugs and alcohol. I mean to say that you could be addicted to chocolates or coffee or Premier League football and would get upset and angry when you don't get your fix. However, the chances of you resorting to violence or crime are very unlikely. Some children are addicted to mobile gaming, such as PUBG and Fortnite, and there have been reported cases of violence. However, those are very rare and extreme. The point I'm making here is that addictions are dangerous, and the pleasure principle is not a glorified one. It shows us how weak we are as human beings to be enslaved by these addictions.

When I discussed earlier how too much of anything is bad for you, I would like to add that moderation is key to controlling harmless addictions such as chocolates and coffee. However, I wouldn't even recommend moderation in drug or alcohol use because that use will become abuse. The pleasure principle applies here because once you start enjoying the high or the buzz, you want to keep enjoying it, and you'll never get sick of it. It is a highly unhealthy habit and a social disease because it tears apart a human being inside out and turns them into an outcast. That person is shunned by his family, friends, and society and could potentially spend most of their lives behind bars or in rehabilitation centers. They end up wasting their entire lives entertaining their addictions instead of living everyday lives and seeking larger ambitions and success, like most of us want to. The pleasure principle is only going to give you more pain than pleasure. There is absolutely no pleasure in this kind of pain and whatever kick anyone gets out of these drugs is only temporary. There is no health benefit to it but only self-destruction.

Living on the Edge

As explained above, most people refer to substance abuse as the illegal use of drugs. However, there's more to it than just consuming illegal drugs. These alter the user's moods, but more than that, they could easily cloud your judgment, distort perceptions, and alter your reaction times, according to the website verynicemd.com. As explained above, these would cause you bodily harm, injuries, and accidents. You may wonder why alcohol is lumped into substance abuse too. The idea is the same in both cases because the pleasure principle is very much a part of drinking as any other drug. Drinking does pretty much the same things mentioned above, such as mood alteration, distorted perceptions, and altered reaction times. You lose control of your thoughts and minds, which is where the word "wasted" comes from. You simply have zero control over your senses, and that's where things get nasty. Alcoholism is a severe problem and has caused so many broken families and unemployment because the alcoholic becomes mentally unstable because he can't control his or her drinking habit.

Defenders of alcohol consumption advocate controlled or social drinking. We mentioned previously about moderation, so generally, moderated drinking is considered permissible and safe. The chances are that moderated drinking can go out of hand very quickly. Alcoholic drinks are available quite conveniently as they don't come under the banner of illegal drugs. This is where we clear the misconception of substance abuse. It is not just limited to illegal drugs but alcoholism and even over-the-counter and prescription medications. We can even include cigarette smoking here because it isn't unlawful for adults to indulge in it. It does have nicotine, which gives the smoker pleasure and becomes a terrible habit. Chain-smoking is known to have so many dangers, especially when it comes to respiratory problems. It is known to cause damage to the lungs and heart diseases and even lung cancer. The effects of smoking are dangerous, yet cigarette packs are available as conveniently as are alcoholic drinks. As long as you are of legal age, you can purchase them. Smoking enthusiasts would claim that they smoke for pleasure and control their habit, but they may not be aware that they could quickly lose control. The withdrawal symptoms of smoking are very high as it is very tough to get off smoking for those who have been doing it for years. Quitting smoking can be a painful experience because the body and mind crave nicotine high. This is why many seasoned smokers have resorted to quitting via consuming nicotine patches, gums, and even e-cigarettes and vapes. Either way, they know that they need to keep nicotine in their bodies, and that's why cigarette smoking is very much substance abuse as much as an illegal drug or alcoholic drink.

More than Designer Drugs

We have discussed above how substance abuse encompasses more than just designer or hard drugs such as alcoholic drinks and cigarette smoking. It goes beyond even these two addictions. While in some parts of the world, such as Amsterdam and American states, medicinal use of marijuana is legal and does even include recreational use. Seventeen states in the US have legalised the recreational use of marijuana. There are many variations of marijuana, such as hash, weed, and more, but whatever you call them, they are very addictive and potentially as dangerous as the other designer drugs. Marijuana may not be as harmful as cocaine, heroin, acid, LSD, and others, but it has similar effects of giving its smokers a high mood alteration, and so on. The pleasure principle applies here, too, because the marijuana smoker wants to continue enjoying the high and would spend insane amounts of money to fuel his or her habit. The medicinal benefits of marijuana are known, but there are dangers when used for recreational purposes. The point I'm making here is that any substance that can give you temporary pleasure, change your mood, lose control of your senses, and such are hazardous, and its abuse can end one's life or put one on the path of self-destruction. Certain designer drugs such as LSD cause hallucinations, so you can imagine the impact of these hallucinations on substance abusers.

We all have heard that anabolic steroids are considered banned "substances" in sports. Their use and abuse have disqualified professionals from their respective sports and ended their careers. Anabolic steroids come under the umbrella of substance abuse because they have the same addictive properties as other designer drugs and can be extremely dangerous. Some have induced "roid rage" in their abusers.

The dangers of substance abuse are known, but it still exists in our society and has endangered our communities despite knowing the facts. Now that we have explained substance abuse in detail, we will look at its impact on individuals and their families in detail in the next chapter.

Chapter 2: Effects and Impact

In the previous chapter, I explained substance abuse of different kinds and how they take addicts and abusers on a path of self-destruction. While the impact is so great that one gets completely dependent on drugs and alcohol, and withdrawal symptoms can drive one mad, the worst is what happens to your loved ones and the community around you. The social costs of addiction and substance abuse are very high as people end up isolating themselves from the rest of the world. Let's take a deeper look at how substance abuse impacts individuals.

Impact on Individuals

Abusers can cause harm to themselves, their family, friends, and colleagues; also, they experience bouts of violence and uncontrollable tempers and lose their social circle and professional path. They isolate themselves and cause self-harm. They misbehave with their friends and family. Also, they experience negative emotions such as anger, fear, rage, despair, and sorrow. We will examine every emotion and see where these come from. When we like something, we want to have more of it. If you have studied economics, you must have known of the law of diminishing returns. If you don't know what it means, it is the idea that the more you consume something, your utility increases until a certain point where it starts to drop. Quite frankly, it means you're bored of it and want to try something new. When that happens, you lose interest and try to find something better to replace your once-enjoyed utility. However, it seems that the law of diminishing returns doesn't apply to substance abuse.

Elements of Chaos

When you consume more drugs and alcohol, you get so addicted that you can't live without it. Your body gets so conditioned by these drugs that it wants to have more of them in the system. It's not the drugs themselves that are the problem but the chemical elements in them. For example, cigarette smoking is considered very harmful and a strong contributor to lung cancer, but it's not the cigarette itself that is the problem. It's the chemical elements inside the cigarette known as nicotine. Nicotine is the addictive chemical that is found in cigarettes that makes you want to keep smoking. This is what leads to chain-smoking, and that's when you get so dependent on cigarettes, you always want to have a pack in hand. You will be craving it several times a day because your body wants that nicotine in the system. We do know that the tar in cigarettes also contains harmful carcinogens. This is why many health experts believe that cigarette smoking is a form of self-harm. You know that it's going to kill you, but you still want to keep smoking cigarettes. That's an irony because one would wonder why you would consume so much of something that would eventually end your life. This would apply to all other drugs and substances, such as alcohol, because they all have harmful effects and endanger your body. Alcohol, for example, could cause liver damage when consumed too much and possibly liver cancer. Consuming too many alcoholic beverages puts you in a drunken state that makes you lose your senses and balance. Alcohol may be addictive, but it is also harmful in other ways, as you have seen. A drunk person can become very vulnerable to being taken advantage of unknowingly. Incidents of date rape have occurred among women as they have gotten way too drunk and then seduced by men without consent. Drunk individuals are also known to have fits of anger and bouts of violence when antagonised or forced to stop drinking. At the same time, these individuals would spend a ton of money purchasing alcoholic beverages and may come under debt or financial distress because they can't kick their habit.

Recreation or a silent killer

Another example of a popular drug is marijuana which is now made legal in 15 American states, but that doesn't mean it is any less harmful. Although it has medicinal effects and has been used for those purposes, it is frequently used as a recreational drug that gets you "high." It elevates your senses, making you feel like you're in a heightened state of joy. It even increases your hunger, and you become dazed and confused. It is reportedly known to kill brain cells, so it's not doing anything good to you if it's just getting you high.

According to drugabuse.gov, "Hallucinogens are a class of drugs that cause hallucinations—profound distortions in a person's perceptions of reality. Hallucinogens can be found in some plants and mushrooms (or their extracts) or can be man-made, and they are commonly divided into two broad categories: classic hallucinogens (such as LSD) and dissociative drugs (such as PCP). When under the influence of either type of drug, people often report rapid, intense emotional swings and seeing images, hearing sounds, and feeling sensations that seem real but are not." Hallucinogens may also include drugs like ecstasy and Psilocybin mushrooms. A lot of these drugs can be very dangerous because the abuser can see things he may be uncomfortable with, and he can't stop because the body keeps craving it. These could possibly lead to suicides because a person may get way too engrossed in the hallucinations. After all, the individual may think all of what he is seeing is true.

Raising Addicts

The harmful effects are apparent, as you can read above. There are even more dangerous drugs we all know of us, such as heroin and cocaine. While the effects on the person are well documented, the impact on society can never be ignored. A person can harm himself mentally, psychologically, physically, and financially. At the same time, it's not just the individual being affected, but also his family, friends, and loved ones. It can lead to broken homes and domestic violence cases. An abuser can turn violent towards his family, and he would be forced into rehab or asked to leave his home because of his mentally unstable condition. The dangers are very apparent because he could harm himself even further and perhaps even take his life. Children raised by parents who are addicted to dangerous drugs and substances could be potentially abusive caregivers, creating a very unhealthy environment at home. Parents could be abusive towards their children, leaving a long-lasting negative impression leading them to develop tons of psychological issues while growing up. We will see how this impacts society and the community because they could become hazards to their community in the next section.

Use and Abuse- Statistically Speaking!

It's pretty clear how bad the impact of substance abuse is, but we need to look deeper into personal impact because that's where it all starts. It then impacts society negatively, but the abuse has to stop from the individual.

According to Surgeongeneral.gov, substance abuse has immediate and direct consequences. The stats reported by them are simply staggering. The report states, "Substance misuse can have immediate, direct consequences for health ranging from effects on heart rate and regulation of body temperature to psychotic episodes, overdose, and death. Many more people now die from alcohol and drug overdoses each year than are killed in automobile accidents. The opioid crisis is fueling this trend, with nearly 30,000 people dying due to an overdose on heroin or prescription opioids in 2014. An additional roughly 20,000 people died as a result of an unintentional overdose of alcohol, cocaine, or non-opioid prescription drugs."

Similarly, there are indirect consequences too as follows:

> "Indirect consequences related to risky behaviours that often accompany alcohol and drug misuse: Alcohol and drug misuse can impair judgment, leading to risky behaviours including driving under the influence (DUI), unprotected sex, and needle/syringe sharing. Driving under the influence of alcohol or drugs contributes to thousands of deaths annually, and 10.6 percent of drivers report engaging in this hazardous behaviour each year. As misuse of some drugs such as prescription opioids progresses, many people seek to intensify the high by injecting them, and sharing of needles among users can result in outbreaks of HIV and hepatitis."

The above statistics don't lie, as substance abuse has added quite a sizeable death toll. If you're under the influence, then you can't function and have a normal life. If you get behind the wheel, you're bound to injure yourself and others. The impact on self and society here is very dangerous. In fact, a lot of innocent lives are at stake, which is a very alarming fact. Let's not even discount the long-term impact on a person's physical and mental health. As reported by the same source, "For example, heavy drinking can lead to hypertension, liver disease, and cancer; regular marijuana use is associated with chronic bronchitis, and use of stimulants such as cocaine can lead to heart disease. In addition, substance misuse during pregnancy can result in long-lasting health effects for the baby, including fetal alcohol spectrum disorders (FASDs), which are estimated to affect as many as 2 to 5 percent of the population, and neonatal abstinence syndrome (NAS); the ongoing opioid crisis has resulted in a five-fold increase in the number of babies who are dependent on opioids at birth."

Substance abusers have no idea how much harm they're inflicting on themselves because they're ruining their bodies to a point where there may be simply no turning back from. As we have noted here, the problems are amplified if they can't afford to fix them. They would be paralysed their entire lives and would just continue to stay this way and eventually die of heart or liver disease or something else related to abuse. They could just crash their car and end up being fatally injured, and there may be collateral damage too. So many innocent lives had become casualties of addicts driving under the influence when they were simply crossing the road. If the driver survives, he may never live with himself knowing he took a life involuntary and could face jail time for it too. Substance abuse should never be taken lightly, and the statistics are evidence of this.

Impact on Society

Though it seems an individual act, substance abusers cause great harm to society. Addicts can turn towards crime and disturb public peace as they lose their employability are considered a threat at the workplace and in public places. The effects of substance abuse include child custody losses, child abuse and neglect, addiction risks for children of drug-addicted parents, domestic disputes, homelessness, poverty, financial healthcare burden, strain on co-workers, increase in imprisonment, and violent crimes in educational institutes.

Life of Crime

Perhaps one of the biggest impacts drug abuse can have on addicts is forcing them into a life of crime as they run out of funds to feed their addiction. They could turn to theft, which means bank robberies, or at other places where they could sell things to get enough funds and purchase their drugs. They would also be mentally unstable, and such people shouldn't be left out in the open. They would never be considered for employment as they would fail the drug tests and perhaps end up behind bars for criminal activities. They could cause chaos in the community due to their violent behaviour. Some shooting sprees have been reportedly related to substance abuse and mental illnesses. While not confirmed, certain mass shootings have been attributed to mental illnesses, such as the Virginia Tech and Connecticut school shootings.

The New Abnormal

These addicts may never be able to have a normal life if they don't overcome their addictions because their entire lives will be centered around feeding their cravings. According to a 1999 report by the Office of Assistant Secretary of Planning and Evaluation at the US Department of Health and Human Services,

“Substance abuse affects and costs the individual, the family, and the community in significant, measurable ways including loss of productivity and unemployability; impairment in physical and mental health; reduced quality of life; increased crime; increased violence; abuse and neglect of children; dependence on non-familial support systems for survival; and expenses for treatment. The physical and mental health and social consequences of alcohol and other drug use by women can seriously affect their lives and those of their families.” Furthermore, a 1996 report by National Center on Addiction and Substance Abuse states that “Not only are women, especially young women, beginning to close the gap between female and male consumption of alcohol and other drugs, they suffer earlier and more serious consequences. Women become intoxicated and addicted more quickly than men and develop related diseases earlier.” As per this report, women have a tendency to get intoxicated and addicted faster than men and also “develop diseases.” Therefore, women should take precautions because they have to take on several roles as adults, including motherhood. Their intoxication and addiction could have a larger negative impact on their children. They may be encouraging their children towards intoxication and addiction too. These children may grow up mentally and emotionally disturbed and could possibly turn towards a life of crime. Whether the parent is male or female, they have a responsibility towards their children and raising them right.

The Cost of Addiction

According to the NDIC, that the annual costs of emergency department visits related to drug abuse are (estimated to be) $161 million, with an additional $5.5 million in costs for those who have to be hospitalised. You can just see how problematic these figures are. Substance abuse takes a mental, physical, and emotional toll on a person, and the monetary cost impacts the healthcare system. In countries where the government provides healthcare, that would be a lot to bear and severely impact the budget. Society pays that price as healthcare is government-provided, but it comes out of your taxes in most such places. There's never a thing such as a free lunch, and everything has a cost. It's costly for prisons to keep substance abusers behind bars, too, which is a federal expense. Similarly, medical treatments, rehabs, and psychiatric treatments don't come free either. If the addicts don't get the proper help because of financial hurdles, they go deep down the rabbit hole and never be able to climb up.

Society's Outcasts

Substance abusers or drug addicts, whatever you call them, become outcasts of society because they have become too toxic for their families. Their addictions have made life hell for their loved ones, so they are forced out. However, the long-term impact of these outcasts can't be ignored. We have seen how much of a danger they can be to society. The statistics here are not surprising, and you can easily look up the damage that substance abuse has caused to society. According to Surgengeneral.gov, long term consequences include but are not limited to reduced productivity, higher health care costs, unintended pregnancies, the spread of infectious disease, drug-related crime, interpersonal violence, stress within families, and many other direct and indirect effects on communities, the economy, and society as a whole. If these people never got into substance abuse or drug addiction, they could have become productive members of their communities and earned an honest living. Chances of a better and healthy life increase ten-fold without all of these harmful drugs in your system. These are poison, and by feeding your body this, you'll be killing it unknowingly. This is something I have touched on a lot in the chapter, and I simply can't stress enough that we, as a society, have to take it seriously and help these people. If we leave them as outcasts, they will only harm themselves and us.

The report by Surgeongeneral.gov tells us things we can simply not ignore. It only makes me depressed, but we have to tackle these issues before becoming a menace to society. The report states as follows:

> "Some of these consequences result from substance use disorders, which occur when a person uses alcohol or drugs to such an extent that it causes clinically significant impairments in health, social functioning, and voluntary control over substance use. The majority of individuals who misuse substances do not develop a substance use disorder. However, roughly one in seven people in the

United States (14.6 percent of the population) are expected to develop a substance use disorder at some point in their lives. A substance use disorder can be diagnosed as mild, moderate, or severe depending on the extent of a person's symptoms. In this *Report*, addiction refers to substance use disorders that can be categorised as severe and are associated with compulsive or uncontrolled use of one or more substances. Addiction is a chronic brain disease that, like milder substance use disorders, has the potential for both recurrence and recovery."

To Quit or Not to Quit

If we read the report, there is one thing it clearly points out regarding substance abuse. It states that substance abuse is a chronic brain disease that has the potential for recurrence and recovery. So, you can recover from it with a lot of effort, but it needs to be controlled. Substance abuse can reoccur if the former addict isn't kept in check. Tendencies and temptations can arise, and it's seen a lot in cases such as when someone is trying to quit smoking. He would successfully quit for weeks to a month but could return because the temptation is still there. The withdrawal symptoms can be too much, and they have to get back to it. The reason is that their health is getting affected by not indulging in smoking because the body needs the nicotine. The former smoker would have to light one up just to keep himself from going berserk. Quitting requires a lot of effort, and we will touch upon this in later chapters, but we can say there are ways to quit smoking that have proven to work. Some people have reported that shifting to nicotine patches, gums, and e-cigarettes/vapes, have helped, but their long-term effectiveness is debatable.

This goes back to the point we discussed in the first chapter about the pleasure principle. The body and mind derive plcasurc from drugs, and these highs and hallucinations are exciting but are more harmful. The pleasure in this pain is irresistible, so no matter what the individual does to get rid of this disease, it keeps coming back like cancer. Not even chemotherapy would help, and there's no pun intended here. This addiction disease is bound to cause even more diseases, and that's when chemotherapy will become a necessity. That's when one will feel very sad that a life has been wasted because the person indulged in addiction and could never quit. The solutions are there, but the costs are high. Therefore, the individual needs all the support, so society is protected from him and his actions.

At the end of the day, drug and substance abuse incurs heavy-duty social and mental costs to the individuals that are involved in them and their loved ones. On the whole, society suffers, and lives get disrupted. You're knocking on the door of death if you indulge in substance abuse because your life may never be the same. The cheap thrills aren't worth the dangers to yourself or your loved ones because they don't deserve this, and you don't need to put yourself through this. The pleasure principle will remain, but we can't stay silent on this matter. These are our people, and they need help. If you are an addict and reading this, you owe it to yourself to take treatment. We will discuss it further in the book, but the time is now to take action. It's best to say no to drugs as that's the only way to protect yourself and your loved ones! The next few chapters will cover more detail on the behavioural impact of substance abuse. It's important to know how an individual's attitude changes in more detail because that's the only way we can understand him and are able to help him. The more aware we are of the metamorphosis people go through while they partake in substance abuse, the better we can explain to our loved ones its dangers. This is the best way we can protect them because knowledge is power. This is why I want you all to read this book and learn from it. In doing so, you can help those who are already impacted by this disease.

Chapter 3: Behavioural Changes

This chapter will cover the behavioural changes in sufferers. We will discuss how adversely these behavioural changes affect how they interact with others, i.e., their family, friends, and colleagues. Further, we will see their typical behaviour and assess their demeanor and attitude as they conduct their daily tasks. Do they stay aloof or become violent when they can't get their fix? All this and more will be included in the chapter.

Over the last two chapters, we have learned about substance abuse and its damaging impact on individuals, their loved ones, and society. If you want to explain it in one sentence, you can say that these dangerous drugs and alcohol can simply hijack your mind and body and make you lose control. The effects are pretty straightforward, and if you know someone who is suffering from substance abuse, you can see that, and the behavourial changes are evident. We will go much deeper to understand these changes in a person, so we know how to help them and recover. I'd like to go back to the pleasure principle concept in this chapter because the person suffering from substance abuse doesn't think he or she is suffering. Unfortunately, they are getting what is known as a "kick" out of it. The pleasure in the pain makes them so numb that their change in behaviour holds no consequence to them. They lose awareness of how such changes impact others around them. These drugs induce a kind of pleasure that keeps them hooked. They lose their sense and mind of what's right, wrong and become socially awkward. There are several ways to tell what's wrong with a person when they suffer from substance abuse. The drugs vary in potency and strength, and the stronger they are, the effects are even worse. If you know someone with a substance abuse problem, look out for the impact we will discuss below.

Social Problems

One of the significant issues that exist in those suffering from substance abuse is social issues. These become socially aloof and distant from their family, loved ones, friends, peers, and work/ school colleagues. They would feel isolated because they feel that their loved ones don't understand or empathise with them. This is a common concern amongst addicts because they see their drugs as a panacea instead of poison. The pleasure principle implies the same, and that's why everything else doesn't matter except the high. You'll see them miss work, school, and other significant life events. Slowly but surely, they will be isolated and withdrawn from activities and daily chores. They will be very secretive about where they spend their time, and most times, they are out to "score" drugs. They will spend time staying awake or sleeping at off-hours until they don't get their fix. These concerns need to be addressed where loved ones should replace the panaceas with affection and support. They can help the sufferers fight the addiction and deal with their anger and mood swings.

Financial Issues

There is no such thing as a free lunch, and that applies to drugs too. Drug peddlers will make their addict buyers pay more money than they can afford, causing them financial problems. Addicts and sufferers will fall into debt traps as they will be looking to borrow money from friends and families to feed their addiction. Otherwise, they may end up in a life of crime. They will constantly be seen running to their kith and kin to get funds and may get angry and violent if they don't get the required funding. They will go knee-deep in debt from loan sharks, and they know they won't be able to pay it off. Hence, their lives will be turned upside down, and they may end up in legal trouble. This means they will be hunted by law enforcement for theft or other illegal activities. Most of these drugs are illegal, and law enforcement won't take kindly to these offenders. If they end up behind bars, they won't be able to pay any fines or bails to get out of jail. This will create even more financial and legal hurdles.

This applies to alcoholics and smokers, too, as they may be seen borrowing money to buy their drinks and cigarettes and getting in debt at the same time. You may even see such absurdities as addicts and sufferers will be stealing valuable household items to fund their addictions. This would mean stealing from spouses, parents, and even children if they're so desperate. Burglary could very well happen, as explained above, and that's when the alarm bells ring. If you notice suspicious behaviour and missing items in the house, your loved one could be suffering from substance abuse. They may get desperate enough to sell their personal belongings to feed their addiction because they aren't thinking at all. Being hostage to drugs and alcohol, they're only looking at ways to get enough cash to pay for their addictions. They will be making several visits to the pawnshop to get rid of jewelry and other valuables. Their financial troubles will keep multiplying unless they kick their habits. No one substance abuse victim suffers similarly. They need to be kept in check because they may end up selling their house just to pay for drugs if things go too far. That is very dangerous, and families and loved ones need to take notice.

Mood Swings and Emotional Problems

The chances are that sufferers of substance abuse will be having plenty of mood swings from anger to frustration to joy because the drugs are controlling their mood. As long as they're getting their fix, they will be very calm and cool. Otherwise, they will have fits of rage and anger, something we touched upon in previous chapters as well. Interpersonal problems and issues will multiply as they will not be able to maintain relationships with their loved ones. If they're married, their marital relationship may fall apart, affecting the lives of any children they have. These issues are going to impact their lives severely, turning them into social outcasts pretty surely.

You'll see the sufferers lose friends because they will become too toxic, and no one prefers such company. Due to their emotional issues, they will find themselves not taking part in any social activity at all. The social and emotional issues pretty much go hand in hand, and these are concerns that we can't afford to ignore. Their minds will be in control of the drugs so much that confusion will almost always set in. You will find them dazed and confused for the most part because they're now hostage to the drugs. They will act selfish, immature, and childish and become way too toxic and disturbing to people around them. As a result, they will be forced out of their jobs as they will be too unsettled to work properly. That will incur the same financial issues we discussed earlier in the chapter.

If you try to confront them for their problems, they will come up with blatant lies, excuses, and justifications that will not make sense at all. They will play the victim game, saying that they are not being understood in society and their families, so they turn to drugs and alcohol. They may keep their addictions secret and pretend to behave normally in front of their family and friends. That brings a more profound issue because you don't know when they will react violently due to their addictions. We need to be more aware of these sufferers because we are too wrapped up in our lives to notice, bringing dangerous precedence. Since every sufferer reacts and acts differently to the drugs, you have to be more vigilant to see their patterns, and that is why we are discussing these behavioural changes in this chapter.

Mood swings, as discussed above, are common, but sometimes you may never notice them at all. They will be delighted one day, disinterested the other day, and most of the time very unfocused and making mistakes in daily activities. For example, driving under the influence is very obvious for us to notice. Not being involved in an activity or outing is not apparent at all. We have to be very mindful of their behaviours.

Lying is something that is a very apparent symptom of drug abuse. This is because sufferers are too afraid to tell others about their problems and will constantly resort to lying. They will be making new friends you never know, and those would be the drug peddlers that we all should watch out for. Their company will not be the most bright or positive because addicts tend to hang out together and also buy from the same people. Whenever these people call the sufferers, they will disappear and go into some other room or turn the phone away. They would constantly act mysterious, and that will be a sign that something is definitely wrong.

You will notice that while abuse victims and sufferers are not candid about their problems, they hide them because of shame. They will be too afraid to be judged as social outcasts, and hence they will resort to being mysterious but never upfront until you notice their behavioural changes. They will feel that they can handle things alone, but unfortunately, they won't be able to do so. They will avoid eye contact, make absurd excuses, and dodge your questions once you confront them. This is a clear sign that they're hiding their addictions due to shame.

Lack of hygiene and cleanliness

Drug addictions and substance abuse are so dangerous that victims will surely be putting them above everything else. You will see this a lot, depending on the severity of the addiction. Health concerns are obvious, but what you'll see is a slow but steady fall from grace, and sometimes it's way too drastic for you to imagine. They're falling since they were addicted. They will not shower regularly and be smelling of alcohol or be dressing shabbily. They will also not be taking care of themselves and being unclean at all times. If you see someone who was very particular about taking care of their health and hygiene suddenly ignoring all of that, that's a sign of substance abuse. These are clear and apparent changes, and if you confront them, they will make excuses that they're not feeling well or just bored with life.

The hygiene issues don't stop there. A lot of times, victims won't be looking clean is because of lack of sleep. They will simply be staying up if their cravings aren't satisfied. Sometimes, they will be way too lazy and sleep a lot more than you expected them to. If they're lazy, they'll shower very little and not end up at their places of work, and that would ultimately lead to their termination. Hence, that would lead to all the financial troubles that we discussed earlier. They will also either lose or gain their appetite, depending on the drugs they're taking. Substance abuse takes over everything from body to mind. They may lose or gain a lot of weight, and those are signs of unhealthy behaviour. You may see the changes and may choose to ignore them as normal but pay heed as it could be substance abuse. These days, victims are very clever at hiding, as we discussed earlier, so they will not be so upfront about their habits and continue to experience physical changes from diet to their appearance and cleanliness.

Physical issues and changes

If the diet and weight changes aren't enough, you'll see a lot worse in a substance abuse victim. Physical changes are a dime a dozen, and that's why it's imperative to take notice of these changes. Some of the signs you should look for physically in an abuse victim are marks on arms caused by needles, which could indicate them abusing heroin or similar drugs, or bloodshot or dilated eyes. The latter is very common among abuse victims as the eyes may get bloodshot, one of them may be more minor, or the other, and speech will be slurred. They will be looking and feeling constantly confused as a result. You'll see changes in skin complexion, too, along with weight loss or gain. If they smoke many cigarettes, they will have damaged teeth, which will be apparent by just looking at them. Cigarette smokers also develop dark lips, which are very common in singling out smokers. Their teeth would get damaged the longer they smoke cigarettes. They will constantly be smelling of smoke too or alcohol if they are drinkers (as discussed earlier). Weakness and fatigue is another sign that they are suffering from substance abuse. These are all the signs that you need to look for and recognise so they can get treatment at the earliest, all before the abuse gets way out of hand.

We have discussed all the more common and uncommon behavioural changes in substance abuse victims and sufferers. As we read ahead, these are things you need to keep in mind because one of us may already be suffering from them. We need to reach out and seek help for these victims, and one of the reasons for writing this book is to give you an idea of how to take care of these victims because they need our love and support to kick their habits. Physical, emotional, and hygiene changes are all things we need to be vigilant of in a substance abuse victim. Furthermore, financial issues are also deep concerns because they could lead to a lot more dangerous habits such as crime, murder, and a life spent behind bars. I know I hadn't mentioned murder earlier, but it is possible, and I don't want to name any examples, but this is something you can expect from victims.

When we discussed violent behaviours above, we didn't touch on this right away because this is very sensitive. If drug addictions get out of hand and take complete control of the victim, a whole lot of personal, physical, and collateral damage can be caused, and innocent lives may be lost.

Dealing with Behavioural Problems

When we read these, we may get scared because if we know someone is suffering, we could easily shun them away, but we don't need to go down that road at all. We have to recognise they are suffering and help them overcome their addictions. We don't need to turn them into outcasts because they could become hazards to society. We will go into the dealing aspect in later chapters, but I just want to touch on the fact that we can't ignore these factors at all. I want to educate you by informing you about these behavioural changes so you can help your loved ones and friends suffering from substance abuse. It's not easy, but together, we can make a difference. Every problem has a solution, and once we have seen the signs, we need to come up with the solution. You can confront the victim and fight them, but that's never the answer. They will react more violently in return, so you need to be careful and very loving.

Love conquers all, so it's important to exercise self-control and calm because you may become violent in response to their violence, which will not help anyone. You have to be very careful in dealing with substance abuse victims because their behaviour is dangerous and may harm you if you resort to violence and anger yourself. They are too much at the mercy of their addictions to care about what you're saying. They will simply ignore you and get even more isolated. Their behaviours will only be more amplified, so you have to be very careful about dealing with these victims. There are many ways to help, which include counselling, medications, and rehab. At the end of the day, each substance abuse is very different from the other and needs to be treated on a case-by-case basis. Therefore, we can't talk about it without digging deep into the individual issues. You have to recognise patterns and behaviours, see what treatment works best, and then go for that. Read on, and you'll see what I feel are the best ways to help substance abuse victims overcome their addictions because it surely deserves a chapter on its own.

Chapter 4: Mental Health

Substance abuse is one disease that could very well lead to mental health illnesses such as depression, insomnia, anxiety, stress, and suicidal tendencies, among others. Various symptoms accompany each condition. Withdrawal symptoms of substance abuse can lead to even more mental health issues when the abuser is craving for his fix and cannot get it. Behavioural changes and relationship issues are all very well tied together when it comes to substance abuse because it impacts our body and mind.

We all know that our minds and bodies are connected like yin and yang. Anything that impacts the mind impacts the body, too, so our body is also healthy if we're mentally at peace. Therefore, mental health is crucial in today's world as we live life in the Fastlane. It gets worse when the stress gets too much, and we look for escape in dangerous and addictive drugs and alcohol, which starts a journey to the dark side. The self-destruction has begun, and we go on a downward spiral where it would be hard to return from. In this chapter, we take a look at the impact of substance abuse on mental health because that's where it all begins. Then, it goes a lot deeper and puts the substance abuse at significant risk, and the worst could be suicide. We touched on these symptoms in previous chapters when we discussed behavioural changes, but we will dive deeper to see the connection between mental health and substance abuse. At the same time, we need to educate others that there are better ways to treat any depression-related symptoms than just get into drugs. A lot is going on in our lives, and if we can manage our schedule and our relationships, we don't need the help of a drug because that's a temporary escape. It only makes you more dependent on it and does not let you free. To understand why all this happens, we need to return to the pleasure principle.

The Pleasure Principle Redux

We have discussed the pleasure principle a few times before in this book because I feel it's the root cause of addiction. Substance abuse starts when you do too much of the same drug or start drinking more than acceptable levels. This is where the pleasure you derive from it increases the more you do, and sometimes diminishing returns don't apply. When we studied economics, we learned diminishing returns, and we concluded that at some time, the utility derived from the consumption of a good or service decreases. You can watch the same movie a few times repeatedly until it gets boring. The same goes for a song, or a snack, or a video game, or anything. Once you reach that saturation point, you decrease less of that good or service. This doesn't apply to substance abuse as the substances you abuse are too powerful and control your mind. They make your mind think that your body can't survive without it despite bringing a lot of pain and trauma to you. You derive pleasure from that pain, and that's when things get very dangerous.

We need to understand that pleasure is just the drug telling you to keep consuming more of it. The pain you are getting from it should be enough to get you off it, but these drugs are way too overpowering. There's a catch here because when we talk about mental health and substance abuse disorders, they are very well connected and correlated. There is a direct proportion, and it goes both ways. You can suffer from mental health issues and start abusing drugs and alcohol to cater to them. You look at them as treatments, and this also applies to painkillers too. Painkiller addiction is hazardous because too much of anything is very bad for you. Anesthetics should be used as advised by the doctors as there can be adverse reactions otherwise. They do help when you're sick but are not used as a recreational drug. According to a report by the National Institute on Drug Abuse,[1]

> "Many individuals who develop substance use disorders (SUD) are also diagnosed with mental disorders and vice versa. Multiple national population surveys have found that about half of those who experience a mental illness during their lives will also experience a substance use disorder and vice versa."

---

[1] *Part 1: The Connection Between Substance Use Disorders and Mental Illness*. (2021, April 13). National Institute on Drug Abuse. https://www.drugabuse.gov/publications/research-reports/common-comorbidities-substance-use-disorders/part-1-connection-between-substance-use-disorders-mental-illness

So, if any of your loved ones, friends, and family suffer from mental health issues, they need to be monitored for indulging in drugs to heal their pain. They need to be told that there is no pleasure derived from these drugs as they would ultimately kill them or wreck their lives. Mental health is an issue that needs to be taken very seriously, and it's up to the NGOs, government, and therapists to be more active in treating these issues. The National Institute on Drug Abuse report cites many mental health disorders that lead to substance abuse problems. Some of these include bipolar disorder, attention-deficit hyperactivity disorder (ADHD), psychotic illness, borderline personality disorder, and antisocial personality disorder.[2] Those suffering from schizophrenia reportedly suffer from substance, alcohol, and tobacco abuse. These aren't disorders you can take lightly as the abuse worsens those suffering from them.

Road to Substance Abuse

We can even include suicidal tendencies, insomnia, anxiety, and stress here. Besides suicidal tendencies, the other mental health issues are pretty common, but that does not in any way mean that you can't suffer from substance abuse if you face any of these issues. There's always a chance that when any stress

---

[2] *Part 1: The Connection Between Substance Use Disorders and Mental Illness*. (2021, April 13). National Institute on Drug Abuse. https://www.drugabuse.gov/publications/research-reports/common-comorbidities-substance-use-disorders/part-1-connection-between-substance-use-disorders-mental-illness

or anxiety gets worse, you may want to indulge in a designer drug for an escape. That then becomes an endless pursuit as you want to get more of it, and you can't get enough of it. It becomes a habit, and then it becomes abuse.

Before moving further, we need to figure out what causes these problems in the first place. There are several reasons why one may become a substance abuse victim, and we'll outline them below.

Unemployment

Unemployment is something we all have experienced at least at one time in our lives if we have chosen to work for others than ourselves. This is a time when you've been laid off, or you have resigned and are looking for new work. There could be a time when you may end up being unemployed for a long time. This could put you in tons of stress and anxiety due to issues faced by yourself and your family trying to cope with daily expenses. You're in a fix because you were living from paycheck to paycheck, and then you're out of a job. At that time, you are stuck in a rut because you're unable to find a job, and the pressure is getting to you. You'll have arguments with your spouse or housemates, and then you may resort to drinking or cigarettes or hard drugs. One drink or puff becomes two, and then it becomes 100 before you know it. The worse outcome would be suicidal tendencies as a result of the drug or alcoholism. You could get sick and suffer physical ailments because the drugs take a toll on your body. This is the time you'd require the support of your friends and family until you secure a job because you could end up getting in worse trouble with drugs and alcohol abuse. That is never the answer, and diving into self-destruction will be its only result.

Business Collapse

Being an entrepreneur is a fantastic feeling because you're in charge of your business and living the life of a boss. You have people answering to you, and you obviously earn more than everyone else in your company. You're living the life, and yourself and your family a great life (if married), and enjoying the best things in life that money can buy. However, we all know that life isn't a bed of roses, and problems are part and parcel of it. Businesses can fail and collapse, resulting in a lot of financial issues. You're the boss but are liable to pay your employees severance as they've been laid off involuntarily. Financial pressures will force you to take on loans, or you may have to file for bankruptcy. Either way, all of this will result in severe stress, anxiety, and depression. The mental impact would be so much more significant as your entire life's dream just collapsed in front of your eyes. You worked very hard to build the business, and having it go down in flames is hard to bear. It's not something for the faint-hearted because a lot of mental and physical issues can result. Heart problems due to stress and anxiety could possibly lead to dependency on painkillers and over-the-counter drugs. It could very much lead to alcoholism, which is when the severe effects of that could result too, such as violent tendencies, loss of mental awareness, being dazed and confused, and so on.

In this situation, you would once again seek support from friends and family, and this could range from emotional to moral support because life can throw curveballs. You just need to be prepared for them, and sometimes they can be very unexpected too. Either way, emotional and moral support is way more helpful than any drug or alcoholic drink. It is vital to seek help in these situations rather than feeling embarrassed and a failure and hide from people. Feelings of isolation do occur, which is why both kinds of support are crucial.

Relationship Problems and Divorce

Here's one more area where a lot of anxiety and stress can happen. Relationships are wonderful when things are going great, and issues occur when things go sour and they are on the verge of collapse. These could be between parents, parents and children, spousal, or just non-marital relationships with a girlfriend or boyfriend. Either way, these relationships are essential because other individuals matter to us in a very special way, and we need to pay heed to how things are going. There will be mood swings, and your relations can be sensitive. The more sensitive they are, the more volatile they can be, and that's when you need to be very careful. If you or the other side doesn't handle it with care, then those can be very flammable and combustible. Relationship problems can be solved via counselling, therapy, or with the intervention of family and friends. When these relationships break, all the issues we have discussed so far can happen, such as stress, anxiety, and depression. When you're facing these issues, you may decide to indulge in substance abuse, and that's where things get worse. We already know what happens next, and when those things do happen, your relationships can get broken for life. Those bridges can be burned and may never be rebuilt.

Here's the thing. If you decide to go down the dark path of drugs and alcohol dependency or what we've been calling abuse so far, it's only going to put you into more trouble than you've been in. Drinking and substance abuse will not repair your relationship with that individual and, in fact, make you look even worse in front of them. This will only lead to unhealthy behaviour and violent outbursts. Suicidal tendencies can occur too, and that won't bring you anything but heartache. You see now that there is zero point in getting into these dangerous habits no matter how bad things get in your relationship. There are ways to fix the issues in your relationships, and those are healthy. As mentioned above, counselling and therapy would prove to be more effective than any hard drug or drink. The mantra you need to follow in times of relationship turmoil is quite simple. It's to fight the problem and not each other. When you practice that, your issues will cease to be issues, and life can get back on track.

Upbringing

The final point we'll discuss in this chapter is about upbringing. This is related to relationship problems, but this goes on a deeper level. When we discussed behavioural problems in previous chapters, we saw how many of these occur due to poor upbringing. Growing up in a broken home or poverty, or even with abusive parents, or those suffering from substance abuse, will automatically result in the child suffering from the same issues and going down the path to self-destruction. Any of these bad parenting and upbringing issues is bound to result in children being raised wrongly or being forced to be on their own. The world out there is not safe for a child already suffering from abuse at home. These are ripe candidates for drug peddlers and would indulge in drugs to escape from years of abuse. This isn't the best idea, to begin with, but how can these children be helped. Educational programmes at school and community awareness can play an essential role in ensuring these children are safe from drugs, cigarettes, and alcohol. It's our responsibility to help these children out and make sure they're safe and healthy. Foster homes can be the right places to keep them if those homes are being run properly. Other than these, there are several ways to help out such children. We have to protect them because these can become community hazards later on in life if they get involved in drugs and become part of gangs, and partake in criminal activities. If they're incarcerated, they would be subjected to further issues behind bars. This would be a never-ending spiral, and the only disaster would ensue. We must not take this lightly and be vigilant with our children and make sure they get the right upbringing. These are the future of our communities and society, so we need to take care of them as our life depends on it.

As with every substance abuse victim, withdrawal symptoms are common, so we need to be diligent in dealing with them. We need to make sure their mental health improves as stress and anxiety are bound to result in withdrawal symptoms. I hope this brings enough awareness of how mental health and substance abuse are linked and how we need to act in situations when the relationship becomes a little too much to handle.

Chapter 5: Treatment Strategies

Whenever we talk about treating someone with a substance abuse problem or drug addiction, we have to look at a few factors. We need to ensure that the treatment is long-lasting because this is not an ordinary disease we're dealing with. This is a dangerous habit that has been repeatedly performed over a long time; therefore, treating is not that simple. It's not as quick and easy as taking a few pills and bedrest, and you're alright. Any kind of treatment for substance abuse has to be effective. When we speak of effectiveness, we mean that the treatment needs to help the substance abuse victims stop indulging in their habit again or even at a later stage in his or her life. It has to allow the victim to reenter society and live a normal life. By normal life, we mean that he or she is able to have a proper job, raise a family, or do whatever they want to do in life. In short, the treatment has to lead them to become productive members of the society they belong to. Any treatment strategy has to be long-term because short-term may not bring about near-permanent results. Perhaps the best way to get out of a drug habit is to have strong willpower. Will power is very necessary because that's the only way a victim would say no whenever they get the craving. There are obviously several ways, and a few popular strategies include therapy, counselling, and rehab clinics. Within therapy and counselling, there are several ways, and when we talk about rehab, we have to look at the quality and success rate of the rehab centre. Therefore, the long-term effectiveness of any treatment strategy is critical in determining whether it's sufficient or not or should be combined with other forms of treatment.

Effectiveness of Treatment Strategies

There are several ways to measure the effectiveness of treatment strategies for substance abuse. The National Institute on Drug Abuse states the principles of effective treatment[3] as follows:

"Based on scientific research since the mid-1970s, the following key principles should form the basis of any effective treatment program:

- Addiction is a complex but treatable disease that affects brain function and behaviour.
- No single treatment is right for everyone.
- People need to have quick access to treatment.
- Effective treatment addresses all of the patient's needs, not just his or her drug use.
- Staying in treatment long enough is critical.
- Counselling and other behavioural therapies are the most commonly used forms of treatment.

---

[3] *Treatment Approaches for Drug Addiction DrugFacts*. (2021, June 10). National Institute on Drug Abuse. https://www.drugabuse.gov/publications/drugfacts/treatment-approaches-drug-addiction

- Medications are often an important part of treatment, especially when combined with behavioural therapies.
- Treatment plans must be reviewed often and modified to fit the patient's changing needs.
- Treatment should address other possible mental disorders.
- Medically assisted detoxification is only the first stage of treatment.
- Treatment doesn't need to be voluntary to be effective.
- Drug use during treatment must be monitored continuously.
- Treatment programs should test patients for HIV/AIDS, hepatitis B and C, tuberculosis, and other infectious diseases as well as teach them about steps they can take to reduce their risk of these illnesses."

When we consider the above principles, we must ensure that any form of treatment applied to substance abuse has to abide by them. As stated above, no one treatment is applicable to all substance abuse victims. It depends on the severity of the substance abuse and its outlying conditions. There is no "one-size-fits-all" treatment, so it's crucial to analyse each patient and customise it according to their symptoms. It has been stated that behavioural therapy and counselling are effective, but those come in several forms. When it comes to behavioural therapy, there are several forms of which the popular ones are Neurolinguistic Programming (NLP), and Cognitive Behavorial Therapy (CBT). They both may or may not include treatments such as timeline therapy and hypnosis, depending on the kind of substance abuse that the patient is involved in. Counselling, otherwise known as talk therapy, is not as extensive as behavioural therapy but may include aspects of NLP and CBT depending on the therapist or counsellor.

Kinds of Treatment Strategies

As discussed above, the more popular and effective strategies include behavioural therapy and counselling. The National Institute on Drug Abuse[4] provides the following five treatment strategies:

- behavioural counselling
- medication
- medical devices and applications used to treat withdrawal symptoms or deliver skills training
- evaluation and treatment for co-occurring mental health issues such as depression and anxiety
- long-term follow-up to prevent relapse

[4] *Treatment Approaches for Drug Addiction DrugFacts*. (2021, June 10). National Institute on Drug Abuse. https://www.drugabuse.gov/publications/drugfacts/treatment-approaches-drug-addiction

When we consider the above five therapies, we can see that they can be used as a combination depending on the kind of substance abuse. There's even the aspect of withdrawal symptoms, and that needs to be controlled as well. In later chapters, we will talk about therapy-based treatments in detail so you will know how hypnosis can help with withdrawal symptoms. If the patient is suffering from other symptoms such as depression, anxiety, or anger issues, those have to be looked after too. So, we can say that there is no one-shot cure for substance abuse because it varies from person to person. At this point, we'll touch on various therapies see how effective they can be in helping someone overcome substance abuse. We will also see how best they can be used on substance abuse victims.

Cognitive Behavioural Therapy (CBT)

Before we explain how CBT works, we need to understand its use and why it's used. According to Mayo Clinic[5], "Cognitive behavioural therapy (CBT) is a common type of talk therapy (psychotherapy). You work with a mental health counsellor (psychotherapist or therapist) in a structured way, attending a limited number of sessions. CBT helps you become aware of inaccurate or negative thinking so you can view challenging situations more clearly and respond to them in a more

---

[5] *Cognitive behavioural therapy - Mayo Clinic.* (2019, March 16). Mayo Clinic. https://www.mayoclinic.org/tests-procedures/cognitive-behavioural-therapy/about/pac-20384610

effective way." This is helpful to help the substance abuse victim think positively and avoid negative thoughts. Positive thinking is an essential aspect of such therapies. Mayo Clinic further states that "CBT can be a very helpful tool — either alone or in combination with other therapies — in treating mental health disorders, such as depression, post-traumatic stress disorder (PTSD) or an eating disorder. But not everyone who benefits from CBT has a mental health condition. CBT can be an effective tool to help anyone learn how to better manage stressful life situations."

American Addiction Centers explains very well how CBT can be applied to substance abuse victims. According to their report, "CBT can be applied in the treatment of many different types of problematic substance use. People treated with CBT techniques learn to recognise and change their maladaptive behaviours. CBT can help people with coping skills, identifying risky situations and what to do about them, and with preventing relapse. This approach is helpful because it can be paired with other techniques. The skills learned through CBT continue to be of benefit long after the initial therapy, and it can be used to treat co-occurring mental or physical health disorders as well."[6]

---

[6] *Substance Abuse Treatment, Techniques & Therapy Programs*. (2021, May 14). American Addiction Centers. https://americanaddictioncenters.org/therapy-treatment

The best part of CBT is that it is one comprehensive long-term treatment that will benefit the patient in living a normal life post-recovery. They will know how to avoid the risk of being addicted again and will make smarter decisions. The point of any therapy is that it needs to do the job it's supposed to do. CBT combined with other strategies is highly effective in that manner, as we have seen from our research.

Contingency Management (CM)

Contingency Management is another form of behavorial therapy suited to specific kinds of substance abuse, such as alcoholism. Opioids, marijuana, and other stimulants, as reported by American Addiction Centers. They are best in encouraging and reinforcing sobriety which is the best way to kick the habit. As we know that people act on incentives, so this method works on that premise entirely. It provides the abuse victim material rewards for overcoming their addiction and staying sober continuously. The best outcome of this method is the removal or reduction of two fatal side effects of kicking the habit- dropping out and relapse. The abuse victim will prefer not to return to his addictions as they see a greater reward in staying sober.

Similarly, they will continue to stick to the therapy because it provides a safer alternative. We have to understand the psychology of the victim based on what we have been discussing all along- the pleasure principle. The high or buss from any drug is what gives the addicting pleasure despite it being painful. You are simply replacing that pleasure with the same high or buzz that's healthy in the form of a material reward. Money, cars, phones, gadgets, or retail therapy work depending on what the person or his family can afford. This is an excellent method and, combined with CBT, can be highly effective.

Motivational Interviewing (MI)

American Addiction Centers' report provides another method to help treat substance abuse, known as motivational interviewing. By its very phrasing, we can see that the point is to motivate the addict to quit. The report states that "MI is a drug addiction treatment method of resolving ambivalence in recovering individuals to allow them to embrace their treatment efforts to best change their problematic substance use behaviour. One benefit of MI is that, despite being facilitated by a therapist, those in recovery develop their own motivation and a plan for change over the course of several sessions, which can provide them with more of a sense of control over the course of their treatment."

The idea behind their definition is quite simple. You're motivating the addict to embrace the treatment and find the courage within them to change. The treatment will vary according to how the addict reacts, so it can be customised according to different situations. He wants to quit smoking for good, so you keep motivating him by giving him solid reasons such as a better and healthier life and fewer chances to be diagnosed with cancer.

Another motivation is to explain to them what is more important such as family, especially children. They can be a great motivator, and therapists will use them frequently to get addicted parents to quit. Any sort of motivation is helpful because the point of the interview is to gain the courage to quit over a period of time. They get more in control of their emotions and behaviours and recognise where they're going wrong.

Dialectal Behavioural Therapy (DBT)

While this method is not explicitly designed for substance abuse but could easily be adapted to tackle it. It works best on personality disorders, especially borderline personality disorder, as reported by American Addiction Centers. DBT, as per the report, works to reduce cravings, help patients avoid situations or opportunities to relapse, assist in giving up actions that reinforce substance use, and learn healthy coping skills. The idea is to tackle the symptoms governing substance abuse, and that's why it's proven to be effective. It works on reducing cravings or highlighting dangers where one could relapse or any injuries. The reason why this works is that substance abuse impacts the victim's personality, so if you treat the personality disorder, you're indirectly dealing with substance abuse, and that is a win-win by all counts.

Rational Emotive Behaviour Therapy (REBT)

Now, think of a therapy that allows a patient to understand their thoughts so they could learn better habits and develop positive thinking, ways, and behaviours. This is what enables them to be emotionally and mentally positive and healthy. This is the basis behind Rational Emotive Behaviour Therapy, as suggested by American Addiction Centers. The idea is simple: it's to inculcate rational thinking from within and to let victims know that external situations do not decipher whether they are happy or not. It's a way for patients to realise what's important, which is similar to DBT in a way. However, this is more of a soul-searching process and recognise that the addictions are not helping them but hurting them. They don't need external stimulants to be happy, so it aims to eradicate the pleasure principle. The pleasure is within you and not in the drug. In my opinion, this is the best method as long as it works because not all patients would be open to such deep soul-searching.

Matrix Model

The Matric Model is another unique therapy-based treatment. As per the American Addiction Centers report, it "employs a combination of various therapeutic techniques and was originally developed for the treatment of individuals with stimulant addictions. Against this backdrop of various techniques, therapists focus on rewarding good behaviours and teaching patients to believe in themselves; self-esteem, dignity, and self-worth. The National Institute on Drug Abuse describes the Matrix Model as mostly focused on 'relapse prevention, family and group therapies, drug education, and self-help participation.'"

Once again, the idea with this treatment is to bring out positive thinking and actions and combines various techniques to achieve that goal; hence it's referred to as the Matrix Model. It's a mix of pretty much everything we have touched on in this section. From providing incentives to positive to teaching them good behaviour to building confidence and boosting their personality and self-esteem, this does everything. This would need to be tailormade to individuals and is considered very effective.

Neurolinguistic Programming (NLP)

Finally, NLP works by getting the patient to develop positive thinking subconsciously to translate into positive actions. It uses goal settings, hypnosis, timeline therapy, triggers, limiting beliefs, and being at cause to improve a person's personality and boost their confidence. It works in the same way as DBT and REBT but finds more common ground with CBT due to both using timeline therapy. NLP is helpful for all individuals and not just addicts but has been reportedly known to help smokers quit habits, so it can help with other forms of substance abuse. We will touch upon it in later chapters, but we find it highly useful because it has been used by the likes of Tony Robbins, who is a highly regarded and famous life coach. NLP practitioners are licensed and become life coaches to their patients and help them set a program to achieve goals and overcome negative emotions, and all of this can be applied to substance abuse. The goal here would be to kick the habit, which is why it's considered very effective.

Good Therapy[7] defines NLP as "a psychological approach that involves analysing strategies used by successful individuals and applying them to reach a personal goal. It relates thoughts, language, and patterns of behaviour learned through experience to specific outcomes." It focuses more on the unconscious mind and how you can respond to different situations by using triggers and interventions. It also helps you to reframe situations from negative to positive. It helps you in setting goals and achieving them. A cool aspect of this is timeline therapy that allows you to go back in the past and remove situations where certain emotions were first noticed and remove them to react to future situations in a better and positive manner. Furthermore, it states that, "Proponents of NLP assume all human action is positive. Therefore, if a plan fails or the unexpected happens, the experience is neither good nor bad—it simply presents more useful information."

Therefore, NLP should prove to work with all kinds of substance abuse victims, and that's where we conclude this chapter. Further chapters will go into different behavioural therapy treatments in more detail, so the more educated you are,

---

[7] GoodTherapy Editor Team. (2018a, December 2). *Neuro-Linguistic Programming (NLP)*. GoodTherapy. https://www.goodtherapy.org/learn-about-therapy/types/neuro-linguistic-programming

the better you can help your family and friends suffering from substance abuse.

Chapter 6: Prevention is Better than Cure

It is always a better idea to solve a problem prior to it becoming one. This chapter will discuss why prevention is always better than cure. Substance abuse can and should be tackled much before someone becomes involved in it. The responsibility falls on the person individually, parents, teachers, and other mentors. Children should be taught about the dangers of substance abuse and drug addiction from early days, so they're aware. They have to be trained to watch out for drug peddlers in school and outside. This way, they shall remain protected from substance abuse from the very beginning. Education and awareness programmes on a regular basis will help young children gain a greater idea and hence will not get involved in substance abuse.

One way to combat any addiction is to prevent it from happening in the first place. When you know that something is dangerous for you, why should you indulge in it in the first place? For example, you know the depth of water in a pool, and it's very deep, why should anyone except for expert swimmers and divers dive into it. The point is that drug addiction or substance abuse, however you may call it, is extremely harmful to a person mentally and physically, and to an extent emotionally. We've discussed the pleasure principle several times in previous chapters, and we have discovered that it's a false feeling of satisfaction and pleasure. The high or buzz you get it is only temporary, and the more you consume of it, the more you'll be dependent on it, so much that you can't carry your lives without consuming it. The withdrawal symptoms are too much for anyone to quit the habit. As we have analysed in previous chapters, we have seen the physical, psychological, and mental impact it has on the substance abuse victim. We also looked at various treatment strategies and their effectiveness in tackling substance abuse and realized that the best treatment is prevention.

If you stay away from drugs, you're only keeping yourself away from the harm these drugs can do to you. Knowing not to play with fire is only going to help you not to get burned. How else can one save themselves from the mental and physical anguish substance abuse provides than not indulging in it at all. There are several ways to do so, which we will discuss in this chapter. Several of those include good parenting, moral support, positive social circle, focus on right hobbies, knowing right from wrong, and living healthy.

Parental Guidance

Parents can play a huge role in educating children about the dangers of substance abuse. There are several ways to do this, and one of them is to have a normal and open conversation with them regardless of age. Show them pictures, and they will see what drugs to watch out for. You can, for example, just talk to them about the dangers of smoking and how it affects one's lungs and causes cancer. Perhaps, you can explain what carcinogens mean and how dangerous they are. You can do skits in front of them, so they know how to approach real-life situations.

Start the Conversation and Earn their Trust

KidsHealth[8] explains a great way to start the conversation with your children.

> "As your kids grow older, you can begin talks with them by asking them what they think about drugs. By asking the questions in a nonjudgmental, open-ended way, you're more likely to get an honest response. Remember to show your kids that you're listening and really paying attention to their concerns and questions."

Your children need to earn your trust, so they do the right thing and say no to drugs. The conversations need to be very friendly and educational, but they can be very creative too. Children admire the creativity in all its manifestations, so singing, dancing, or other ways such as taking part in some play would also help. You could plan activities with other parents and have other children assemble and learn about the dangers of drugs.

The language needs to be very easy to understand, and you have to be very approachable to them because this is a serious concern. You could give examples of people who suffered due to

---

[8] *Talking to Your Child About Drugs (for Parents) - Nemours KidsHealth.* (n.d.). KidsHealth from Nemours. Retrieved July 8, 2021, from https://kidshealth.org/en/parents/talk-about-drugs.html

substance abuse. This is a way for the children to learn and educate their peers so everyone benefits. The idea is to educate children about practicing good health and how one should avoid being pulled into taking drugs. The idea is to teach them to say no because these children are very impressionable and gullible. The more educated they are, the better it is for them. They need to be properly trained by the right influencers in their lives, and parents are the best influencers and teachers. Make sure you win your children over by having confidence in them and their abilities. You need to back them up when they make mistakes and be approving of them. It would be best if you got them to talk about eating and living healthy and inculcating discipline in their lives.

Mix Things Up

You could mix things up by mentioning things like how sugary drinks and chocolates can be bad for you if consumed too much and then talk about how drugs are ten times more dangerous. One major lesson to teach children is never to talk to strangers no matter how friendly they appear. They should never talk to them or accept any gifts or money from them in any way. You may think of taking them to a hospital to show how badly smoking and other drugs impact people but make sure they're old enough to take it well. Otherwise, it will leave permanent scars on them and would scare them too much. It's very much akin to showing young children a very scary movie they would have nightmares of. According to Babycenter, "t's not necessary to teach your 5-year-old the specific names of illicit drugs. Instead, focus on things she already knows are bad, like cigarettes. 'You remember how we said smoking hurts your body? Well, there's a drug in cigarettes that makes them bad for you, and sometimes people use other drugs that are bad for you, too.' She can also learn that medicines are drugs and that medicines can hurt her if she takes them in the wrong amounts or at the wrong times."

Best Friends

When speaking with children, you need to let them know that they can talk to you about anything and everything. You need to become their friend first so they can trust you enough. This builds on what we discussed earlier about earning children's trust and confidence. You need to become their best friend, mentor, confidante, and more. You can approach every problem and issue with thought and care, so the time they reach a certain age, like middle school, they could have serious conversations with you about drug use and abuse. By that time, they would simply trust you enough by having a heart-to-heart conversation with them. Trust can go a long way in establishing a healthy relationship between parents and children. This way, your child will make sure to talk to you before indulging in any habit or hobby because they will know what is best for them.

Health is Wealth

Young children are very excited when it comes to candy, and parents would often bribe them to be good in exchange for them. However, they can pose a serious health hazard if consumed in large amounts. You can approach the issue in two ways. The first way is to let them know with authority that they can't consume candy at a certain time, like before dinnertime. The authoritarian approach is not helpful because they will only get upset and throw a tantrum. I would highlight the television show "Full House" here in which the father Danny Tanner, played by comedian Bob Saget, becomes the best friend and mentor to his three girls and talks to them very warmly about all issues they face in their lives. It's one of the best shows about good parenting.

The second approach is what I'd call the Danny Tanner approach because playing the boss won't help your cause at all. Your children will detest you and not listen to you at all. The opposite approach taught by Danny Tanner is to educate them about the dangers of overconsumption of candy. As explained by Babycenter, "explain your rule – even if you feel like you explain it every single day – and make sure your answer focuses on health and nutrition: 'Why can't you have more candy right now? Because too much sugar is bad for your body. It won't make you healthy and strong, but it will tire you out and hurt your teeth. And if you eat a lot of sugary snacks before dinner, you won't have room in your tummy for healthy foods.' If you've taught her the importance of taking care of her body all along, when she's older, she'll be less likely to want to risk drug use." They'll realize that you're teaching them healthy habits for their best interest, and that's why the Danny Tanner approach works the most.

Just Say No!

When children are at least five years, they may not be very cooperative, so they need to be handled with care. They have to deal with so much peer pressure as they grow older in their pre-teen and teenage years because that's when drug use experimentation is at its highest. The training needs to start from the earliest age but strategically and carefully. If they, at that age, are able to explain their views properly, then they know better to stay away from drugs as they grow older. Whenever these growing children approach you and have questions and state their opinions, you need to be excellent listeners. The first rule of counselling is to be a good listener. Listen attentively and respectfully, and they'll appreciate that a lot. If you keep questioning them, they will only rebel against you, and that's not a good sign. Arguments and debates will amount to nothing, but distrust and that will turn them against you. This is when drug peddlers strike and gain children's confidence by pretending to be their friends and well-wishers and trust them over you.

You need to sit down with them and have long and deep conversations, and then they will listen and understand your point of view. Respectfully agree to disagree with them because they will understand that you at least respect their opinions even if you aren't on the same page with them. You just need to explain where you come from, and you're only helping them stay away from the dangers of drugs and will more than likely be wary of them and learn to say no to them.

Educational Programmes

The school can make a huge difference in children's lives. Early school education can revolve around teaching children many healthy habits in a very effective and friendly manner via skits, plays, and talks with doctors. PowerPoint presentations can help too because children learn better with animation and graphics. Kid-friendly cartoons can be a great way to teach children how to learn about drugs.

Since day one, they need to be taught about the dangers of drugs, so they are conditioned to know that drugs are dangerous. Teachers can work with parents and develop programmes about how drugs can negatively impact their lives. They can be given homework assignments on how drugs can harm them or how more intake of them would lead them towards addictions. There could be substance abuse classes for every grade they get promoted to. According to their grade level, the children could be taught things as they grow and move from primary to secondary education. The awareness programmes could be tailormade to each grade, and then there could be classroom exercises for homework. Examinations, assessments, and quizzes could supplement that to test their knowledge on drugs. Special videos could be screened in classrooms where they could learn about drugs more effectively. There could be monthly visits by doctors, and they could talk to students directly and answer any questions they have about substance abuse. It could be ranging from the dangers of smoking, marijuana, cocaine, heroin, LSD, and other dangerous drugs. These could be complemented by teaching them healthy eating habits. There could be periodic health weeks where children would participate in various activities to learn about the dangers of drug abuse. They would present skits, plays, and talks about drugs so other children could learn. Any interactive way to teach children about drugs is extremely helpful.

D.A.R.E

In the United States, programmes such as D.A.R.E have been in practice since 1983 and are led by the local police department to educate children about the dangers of drug abuse. D.A.R.E stands for Drug Abuse Rehabilitation and Education.

According to D.A.R.E's official website[9]:

> "This year millions of school children around the world will benefit from D.A.R.E. (Drug Abuse Resistance Education), the highly acclaimed program that gives kids

[9] *About | D.A.R.E. America*. (n.d.). D.A.R.E. Retrieved July 8, 2021, from https://dare.org/about/

the skills they need to avoid involvement in drugs, gangs, and violence.

D.A.R.E. was founded in 1983 and has proven so successful that it has been implemented in thousands of schools throughout the United States and many other countries.

D.A.R.E. is a police officer-led series of classroom lessons that teaches children from kindergarten through 12th grade how to resist peer pressure and live productive drug and violence-free lives."

Such programmes catered to students at all grade levels are deemed highly effective and should be encouraged at a global level, as demonstrated above.

Do the Right Thing

Good parenting and educational programmes can go a long way in ensuring our children are made aware of substance abuse and its dangers from a very early age. This is very important because these children could become role models to their peers and become positive parents when they become ones too. For this very reason, we must be very responsible as parents because our children look up to us. We must learn that with great power comes great responsibility. Similarly, with great responsibility comes great power, too, so we must use both of these properly. We need to ensure our children grow up learning healthy habits and stay away from drugs, so they don't have to resort to any treatments we discussed in previous chapters. We also have to ensure they stay away, so none of the behavioural changes that could turn their lives upside down don't happen to them.

At the end of the day, these children are our future, so we need to educate them, and it starts with us as parents and then goes hand in hand with schools such as the D.A.R.E programme in America.

Chapter 7: The Right Environment

As stated in the previous chapter, the right education and awareness on substance abuse have to start at home with children at a very young age. At the same time, it's also crucial to raise children in the right environment. Children being raised in broken homes or having parents who are addicts can have a severe and negative impact on them. As a result, they could be forced to turn to substance abuse. This chapter will discuss how parents should raise their children and instill values that will help them stay free from substance abuse growing up.

Growing children are perhaps most vulnerable to fall into the trap of substance abuse. This is why it's imperative to ensure they're being raised in the right environment. Firstly, we need to constitute what is meant by the right environment. Children growing in broken homes or substance abuse victims will be severely impacted. They will have to face a lot of negativity from parents, feel neglected and unloved. Children spend more time at home than at school, so it's important that they are treated well there and learn the right lessons from their parents.

Raising children right is the responsibility of every parent, which would include creating a safe and open environment initially. Once you set that foundation, you set the stage for children to be nurtured and cared for in the right way. They'll earn your trust and will listen to you and learn from you.

Safe and Open Environment

Every child prefers to feel safe, stable, and secure at their home. Their home is their nest because they feel protected with their parents. For example, when a child goes to school for the first time and leaves their mother, they tend to cry and become difficult. There's a lot of separation anxiety because they're leaving their safe space or, in other words, their comfort zone. The keyword here is comfort, and a child needs to feel comfortable, and when the parents make that happen, the child gets very attached to that space. It's also very natural for children to be attached to mothers at an early age over the fathers. Mothers give birth to children and spend more time raising them, and hence there is an immediate and strong connection.

By being in a secure and safe space, children will experience positive personal growth and higher chances of success in life. They will feel a lot stronger emotionally and will be able to face challenges better than others. At the end of the day, it's about positivity. Throughout this book, we've focused on the dangers of substance abuse, how it impacts personality, and treatment strategies. We also talked about how prevention is better than cure. This last point is so very important, and I can't stress this enough that our children are our lifelines, and we need to take care of them so that they don't go to the wrong people to get that care and support. Those individuals could put them on the path of substance abuse, and it's a nasty place to come back from. We had talked about how to communicate with our children effectively and explained the negative impact of drugs on them.

It's not just about having these Danny Tanner-style talks with them, but giving them the care, love, support, and protection that they deserve. At the same time, you, as a parent, would want to become a very strong and positive role model for them. Children are directly influenced by their parents, so it's imperative that parents set the right example for them. That's how the upbringing starts. If you as a parent stay away from substance abuse, then it is expected that your child will also not fall into that trap. If you smoke or drink, then it would be advisable to control those habits or not do it in front of children until they reach adulthood.

You need to provide them the ultimate security, and that includes physical, social, and emotional security. As discussed in the previous chapter, you need to create that connection with your children, so they confide in you. You need to become their mentors, guides, counsellors, and best friends. This is perhaps the best way around it. When they make mistakes, you don't need to be so harsh with them. You just need to explain their mistakes and perhaps give them a milder punishment if needed. It could be some form of community service or some other activity that could give them some positive reinforcement.

Best Friends

Now that we have established that you, as a parent, need to be your child's best friend. This is a journey in itself that starts from the day your child is born. You have to ensure that there is a meaningful relationship and bond between parent and child. The first step is to give them a safe space they call home. This comes under the umbrella of physical safety, neglect-free, and an abuse-free home. This means that you have to attend to your child's needs at all times and not be abusive towards them. You don't want to push them away from you by becoming hostile towards them.

You need to safeguard their physical health by keeping them away from dangers that could arise in the household. These would include but are not limited to open windows, electric cords and outlets, hot surfaces, toxic substances, sharp objects, and weapons. You need to teach your child not to play with fire and how to respond to emergencies when they take place.

Bare Necessities

It is very important to provide food, clothing, and shelter for your child, but childcare goes a lot more than that. You need to go above and beyond the call of duty. You can't overlook the essentials, but you can't limit yourself to that either. Every little effort we make to keep our children happy goes a long way. It all boils down to keeping the home environment open, safe, secure, stable, and positive.

We do know that the world is very cruel, and it is our responsibility to make sure we keep our children safe from that. They will interact with different people and sometimes the wrong crowd. We need to let them know they could come to us for anything. See, we need to prepare them to take on challenges in life, so we need to strengthen them emotionally and morally. There are several ways to make this happen, of which the most important is not to show anger.

Anger is dangerous, and by showing that to your child, you will push him away and make them feel unsafe around you. He would not open up to you and constantly feel scared. Your child may inversely become hostile towards you, and that is never a good sign. You don't want your child to show resentment and anger towards you because you have to understand one thing very clearly. Children at a young age are very sensitive, so they will react and will show their feelings openly. Tantrums and mood swings are part and parcel of being a child, and those can be managed with love and support. Tough love can work, too, by rewarding them for completing certain chores and errands.

The important thing children need to know is that we are all human beings, and none of us is perfect. We are all capable of making mistakes, and we will do them from time to time. We need to let our children know that they can make mistakes too, and there is nothing wrong with that. The best thing is to learn from these mistakes. What you can do is let them know what lessons they can learn from them. It all boils down to communication. Communication is king, and we need to send the right messages to our children. We need to tell them that they should not feel ashamed at all for messing up a few times. They need to accept and own their mistakes and remember not to repeat them. If they do, that's fine too because we can remind them how to avoid them. Shouting to them is a complete 'no-no,' and it comes under anger. Soft demeanor and caring words have a much greater impact.

There will be several times when children are going to act out of line and become very difficult. You will tend to lose your patience, but you have to maintain your cool at all times. You need to show emotional maturity and not let the children's tantrums get to you. You need to explain that tantrum isn't the way to get things. Good behaviour and listening to you will get them what they wish. This way, they'll learn to earn their wants instead of being spoilt rotten. You can sit down with them and have heart-to-heart talks. Understand what they are asking you and teach them strong moral lessons from your life and have them ask you questions. You can always quiz them on what they have learned from you. The best thing you can do is spend quality time with them by taking them out to the movies or to their favorite restaurant or an amusement park. You can play video games and board games at home. The Nintendo Wii is a solid choice for a family gaming console that can get families together for some fun and quality time.

Nothing but a Good Time

One thing we must realise is that our children are what they are- children. These are children who enjoy running wild and free. They want to enjoy their playtime. They want to hang out with other children and have fun. They need to explore their creativity and, best of all, have a good time. There are times we fondly remember our childhood because those were simple and innocent times when we had zero responsibility. We could do whatever we wanted and be free as a bird. It was most definitely us having nothing but a good time. The right environment demands children to have fun with their siblings, friends, cousins, and us. Everyone wants to let loose once in a while, and children deserve it more than ever because it's their age to have fun. They could use their time constructively by going to summer school and learning arts, crafts, music, or anything they fancy. They could participate in sports and have a blast that way too. There are plenty of ways children can be children as long as we let them be.

We should allow our children to be super creative and discover their hidden talents. This way, they'll harness them and find their passions. You can spend time with your children too and have a good time because that will increase your friendship with them. You can go to the park and play sports together or just go out camping, or any outdoor activity.

Detox

Since this book is centred around substance abuse, it is imperative that the right environment is completely detoxed of all substances and alcohol. If you want to set an example for your children, then you need to work on not be entangled in the web of substance abuse. Practice restraint and tolerance and let them know that it's harmful to them. Avoid fights and arguments with your spouse because that would impact the children very negatively.

The idea is to provide an environment that's completely substance-free. If we want to raise our children to be role models to their children, the work has to start from now. The previous chapter emphasized a lot on how they should be educated about substance abuse, but we also need to play our part in ensuring we stay away from it too. The hard work starts at home, and then the children will avoid it at school and growing up. Create a peaceful and non-toxic environment at home, and your children will grow up healthy and strong.

Many of us may have bad habits we can't get rid of, but at least we should control them as much as possible. If we consider the impact it can have on our children, then we will avoid these dangerous habits because you can't create a healthy environment at home otherwise.

If you're a cigarette smoker, avoid smoking in front of children and make attempts to quit smoking because you don't want them to become smokers. Stay away from marijuana because while it has medicinal uses, it's not supposed to be used for fun or to get high. It has adverse effects, too, so one should stay away. You don't want anything killing off your brain cells or theirs.

Similarly, stay off the wine, vodka, whisky, beer, or any alcoholic drink as much as you can except socially and in moderation if needed. That's the best way to go about it because your children could take up drinking after you. It will never be a good idea to encourage them to harm themselves.

Love can Move Mountains.

At the end of the day, our children want nothing more from us other than to love them wholeheartedly. The right environment starts with love, and these are our flesh and blood. They are our future, and we must protect them, and we have a lot of work to do when it comes to raising them right. As you have seen in this chapter, it's very much doable. If you take the right measures, you will see great results.

As we look ahead to the following chapters, you must keep in mind that the nurturing needs to start from the first day. We will look at how to raise our children as they grow older in the next few chapters, and that is a major aspect of this book. The idea is to make sure that our children stay safe and protected because they get the most hurt by our unhealthy choices, and if we live in a home where we are victims of substance abuse, we aren't setting the right example at all.

Before we move on to the next chapter, I must tell you that love can move mountains. It can help our children become great people because we are great people. It's the most incredible feeling to love and be loved. Love your children so they can love you back!

Chapter 8: Staying in School- The Teenage Years

It is extremely important for children growing up to stay focused on academics and cocurricular activities in school. They must be regularly involved in sports and athletics to stay away from the wrong crowd. A healthy environment at home and school will enable them to become good students and athletes and stay away from substance abuse. This needs to continue until they get into the university age.

As parents, it is our responsibility to ensure that our children stay away from drugs and other substances throughout their growing years, especially during their teens. Since they are very impressionable at this time, they can easily be influenced by the wrong people. You don't want your children to be miseducated and misguided by the wrong people. This is also the time they're going through several hormonal and physical changes, and emotions are everywhere. They will get into relationships, start dating, and will need to deal with peer pressure, possible bullying, and self-esteem issues. This is a very pivotal time in a teen's life.

There are several ways we can help our growing teenagers to stay away from substance abuse and live healthy lives as adults. We will list them below in this chapter to help you understand and teach them to your teenagers.

Positive Role Model!

It's about maintaining a positive attitude so that it rubs off on your children. Since children model parents' behaviour, it's important to make sure we abstain from substances such as alcohol, cigarettes, and hard drugs. Therefore, you have to keep a substance-free environment at home to let them know that these substances are off limits, and they must avoid them. If we do indulge in drinking, it should not be in front of them or during festive occasions where they're aware that it's only for adults and not children.

They need to learn how to enjoy life without getting involved in bad habits, and you can guide them to do so. Make sure you practice what you preach. You should help them handle issues that don't involve drugs or alcohol.

Holding Hands

When we were growing up, we looked towards our parents and appreciated their involvement in our lives. We liked it when they played sports with us or video games or anything that was healthy and did not make us unfit. We need to do the same with our teens because they would expect us to be their best friends. We need to hold their hands and walk them through the challenges of life to help them prepare to handle the challenges when they are grownups. The lessons they learn now will help them much later in life. Therefore, it's absolutely very necessary to spend quality time with our teenage children.

We should make it a point to have at least two meals a day with them, which could be breakfast and dinner because if we're working, we won't be home for lunch, but we could still make time for breakfast.

The best way is to schedule a part of the day to spend time with them, and this should be a regular habit. This will increase the bond and trust you have with your child. It could be anything from sports to gaming to even helping them with homework. The more time you spend with them, the closer you'll get, and that will make them listen to you more.

Another way to be more involved in your teens' lives is to show interest in their hobbies and activities. We've discussed gaming, sports, athletics, reading, and other activities; now, you just find something of common interest and spend time doing that. If there is no common interest, work to develop one because that's one way to be their best friends. The idea is to develop a very strong bond with them, so they develop healthy habits and become responsible adults.

One more thing you can do is to have daily conversations and know the people they hang out with. It could be their friends, colleagues, peers, or anyone, so you know where they are at all times. You're able to reduce their risk by knowing who they talk to and hang out with and keep tabs on them. You don't have to police them but let them know you're always available for them. They also will be careful knowing where they are and who they're with because they know you have eyes on them. At the same time, they'll feel secure that they have your backing.

As stated previously, teenagers are very impressionable and can be influenced easily. Pop culture and trends dictate the way they act and behave, and they tend to follow their peers. If their peers are into drugs, they could fall into that trap too. This is why it's important to know their friends and make sure they're making friends with the right people. The best way to do it is to invite their friends over to get to know them better. You could perhaps exchange niceties with them whenever you pick your teens up from school or after their extra-curricular activities are over.

Whatever way you can get to talk to their friends, it will help you gauge if they're a positive or negative influence on them. If they are a negative influence, you need to figure out what makes them such from their parents or guardians. You can find out where they live and their situation. That way, you know more about them and explain to your teens how to help them or perhaps break ties if that friendship will push them towards substance abuse. You need to let them know that they are looking after their best interest. Forming a support network with your friends' parents will help you all keep an eye on your teens and protect them.

These days, your teens will be connected to various social media platforms and could connect with strangers, and those could lure them towards substance abuse. You need to keep tabs on their social media accounts to ensure they're safe from cybercrime or making the wrong friends.

Parental Guidance - 13

The one way to make your teenager understand moral values and develop a sense of ethics is to discuss rules and expectations that ultimately lead to learning discipline, which is important at every age. If your teens learn it earlier on, they would be conditioned to be disciplined later on when they're not holding your hands.

You shouldn't impose rules on them like you're a martial law administrator or an army general. You should discuss the rules with them to let them know that there are consequences to breaking them. They need to know that this is training for their adult life so they can be responsible for the decisions they make. It's adulting 101, so they can be better equipped to handle real-life problems on their own. The rules need to be explained with degrees of consequences with initial warnings and then harsher punishments but nothing that is cruel. They need to know why the rules are important and how they will help them and not hurt them. You don't want your teenagers to rebel against you, so make sure to speak to them on their terms as their friends and not a boss.

While you aim to be their friend, you should never forget your role as their parent, and they should respect you at all times. You can reward them for sticking to the rules and give them cheat days to have fun, and this they will learn how to act on incentives. Positive reinforcement and motivation are key ingredients to build a stronger relationship with your teens.

Always talk to them about your experiences growing up and how you followed the advice of your parents, and how it benefited you. You should explain to them, and I highly recommend that these stories revolve around them, knowing that you were once at their age and how you were disciplined and followed the rules. You can say that your parents also rewarded you for good behaviour and that could range from material rewards to a boost in allowance or a car at the age of 18. For example, you can gift your teenager the latest PlayStation console for developing a disciplined schedule at home and at school for three consecutive months, and that includes making the right friends too.

We live in societies where drinking alcoholic beverages is considered a normal thing, and our teenagers need to know the same too. It is okay, as adults, to drink in moderation, but they don't lose self-control. There are age limits set for a reason because you have to be a certain age to handle the effects of alcohol and know when to stop drinking. If you become over-smart by introducing your teenage children to alcoholic beverages at an earlier age than what the local laws allow, you're going to drive them towards drinking as a minor. You would be pushing them towards substance abuse and alcoholism because you'd make them very curious to know why drinking should be done in moderation.

It would be akin to digging your own grave and burying yourself and your teenager in it. You don't want to become friends with them by making them your drinking buddies. That's not going to help them stay away from alcoholism or substance abuse. It's highly likely that they've been explained the harms of alcoholic beverages at school already. You can simply reinforce those teaching and ensure they don't drink until they're of legal age. Even then, they need to know their limits.

They need to be educated as they grow old about the dangers of alcoholism. When they run off to university, they will be on their own and be exposed to all kinds of positive and negative influences. Alcohol is one of them as their peers would start drinking, and they would want to as well. Therefore, it's important for them to know their limits so they can drink responsibly. They must know how drinking could impair their vision and senses, so they should never drink and drive. Driving under the influence is a crime, and they could face jail time if they do that. This is where following rules will help, as you can let them know it's a simulation of following real-life laws and how breaking them could land them in trouble. The better they understand self-discipline now, the better their chances are having a successful adult life and becoming responsible parents themselves.

Quality Time

These days, time is a real luxury. We live in a fast-paced world where we spend more time on our phones and tablets than each other. Children grow old playing games on their smartphones and have no time to spend with their parents. Similarly, parents are too wrapped up in their lives to be bothered about their teenagers' whereabouts. If you don't have time, then you have to make time and let your teenagers know the same. There is a time to go to school, do homework, play sports and games on smartphones, but children need to know that their parents also require their time. Similarly, parents have to remind their children that they are always available for them no matter what.

We must realize that our teenagers are our priority and everything else from work to social life comes later. We mentioned the Danny Tanner style of parenting from the hit 1980s-90s television show Full House in the previous chapter. In that show, Bob Saget, as single dad Danny Tanner, would balance his other duties with parenting with the help of his brother-in-law Jessie Katsopolis (John Stamos) and Joey (Joseph) Gladstone (Dave Coulier). These three men took responsibility for co-parenting three growing daughters, and the series saw all three grow from young children into teenagers. Danny, Joey, and Jessie would always have time for these girls and sit with them to talk out their problems. They would become best friends. This tradition continued in the later sequel series Fuller House where the two elder Tanner daughters were grown adults and had their families and would incorporate their father's parenting style with their children.

The idea here is to make your child know that they don't need anyone else to counsel them or guide them but you. Parents are the best counsellors, and we need to take responsibility. You need to be a great listener, and Danny Tanner is an excellent example of that. You don't need to sound too authoritative with your teenager but talk on their terms and at their level.

You have to be genuinely sincere and supportive to help them open up to you and discuss their concerns. You have to patiently listen to them and air out your support and guidance, and let them know they have your back. They should know you're always available to listen to their concerns. Try not to show anger but be gentle. You can be polite and gentle but firm when explaining things, so they do understand yet respect you as parents. Even though you want to be their best friend, they need to know you are also their parents and guardians. Your cellphone numbers need to be on their speed dial, and they should always contact you in case of any emergency, whether they're at school, a friend's home, or at a party.

Make sure to form a support network with other parents at their school that also includes those of their friends. You need to have the numbers of their friends so you can contact them in case your teenagers are unreachable. Many issues can be resolved this way, and I can't stress enough that you have to give your teens all the time in their world. This is the age they look up to you the most, so be there for them, and they'll definitely thank you later.

The Living Years

We are living in critical times, and our children are our biggest and most important asset. It is our responsibility to take care of them fully and ensure they don't get influenced by negative people. We need to be there for them at all times and also keep an eye on them as they grow older in their teenage years. It is a time where they're always impressionable and can be easily influenced. Social media can mislead our teenagers into bad habits and in touch with the wrong crowd if not monitored. We must own our responsibility and be their best friends but keep reminding them that we are also their parents and guardians at the same time. Hopefully, all of these things can help us keep our teenagers safe from the dangers of substance abuse.

The best way to do it is to practice what you preach and ensure you don't indulge in substance abuse either and practice moderation when it comes to alcohol. We need to teach them the right habits with rules and consequences so they are prepared to face real life responsibly as adults. We need to teach them responsibility by being responsible ourselves, too; it all goes hand-in-hand. The right parenting will only help our teenagers in the long run and live safe and healthy as adults.

Chapter 9: Staying Safe as Young Adults

Growing up is a rite of passage, and as we grow older, we need to be careful with our dietary habits and fitness level. These are the years where we can keep ourselves fit and healthy. When we become adults, we finally taste some responsibility. University life is the first step to independence and responsibility. It's the precursor to the real world and parenthood.

As we discussed in previous chapters, our children need to stay substance-free from childhood to late teens. More importantly, adulthood is also a time when they could be even more impressionable than when they were growing up. We all have been through these stages, and we know that our children will leave us for university, and our influence on them will gradually decline. However, it is very important that all the teaching, development, and training in school and at home stays with them during university life. It is our hope that it sticks, but this chapter serves a different purpose. Being parents, you could use this as part of your pep talk with your young adult before he goes off to university. It is important that they understand how to stay safe. We can't be with them all our lives because they will eventually become working adults and parents. We can set the right example for them so they can learn to stay safe.

New Horizons

University brings about new horizons to young adults and with that brings new experiences. These are essential in helping them learn the ropes of what it means to be an adult. At university, they are on their own, living independently without parental supervision and making decisions. They have to take care of their needs and expenses. They may be supported completely or partially by parents but are pretty much independent. University is, in fact, a stepping stone to proper adulthood.

It is here that they will make new friends, get into relationships with the opposite sex, and perhaps even do part-time or full-time work along with their studies. They will have exposure to so many different people, which could either negatively or positively influence them. With negative, we do mean that these people could push them towards substance abuse and alcoholism. This is something very necessary to stay away from. When young adults are free from the control of their parents, they can be peer-pressured into anything, and parties in universities are events where this is pretty much a norm.

Party all the Time

The worst thing that can happen to young adults at university is alcoholism. Overconsumption of alcoholic beverages can lead to dependence. It is extremely hard to get out of this rut as they want to party hard with their friends and have a good time.

Getting wasted is considered cool, which basically means getting so intoxicated that you lose a complete sense of self-control. One can become animalistic in nature and can indulge in anything such as getting into a horrible accident, throwing up, and unprotected sex. It's advised to drink in moderation but not overdo it because it will defeat the purpose of having a good time. Partying hard doesn't mean indulging in alcoholic beverages so much that you put yourself and others at risk.

When you're in university, it's hard to stay sober all the time, but drinking in moderation or just socially helps. Just drink enough, so you have some self-control and stay in your senses. Have a friend all the time to help you control your drinking and make sure you don't do too much. A "buddy" or "mate" system helps you keep yourself and your mate in check so you can drink moderately and have a good time. It's not possible to completely abstain from it as you're always under peer pressure to indulge at parties and social events.

The best thing moderation can do for you is condition you for your future life when you start working, get married, and have a family. At those times, you will want to be in more control of your senses as alcoholism will have a negative impact on your personal, professional, and family life. Your friends may not want to be around you too much. At university, it will hamper your ability to succeed in classes and get good grades. Anything in overabundance should be avoided, and that's why it is said that too much of anything is bad for you.

Just Say No!

You must have heard or seen ads on radio and television about saying no to drugs. You've seen it in comic books, too, with your favourite characters stopping you from taking drugs. Remember the "Got Milk" ads with your favourite actors, sportsmen, and other celebrities? These ads were a clever way to encourage people to have milk. Similarly, drugs should be avoided because of the impact they have on addicts. It starts with one or two puffs, drags, or jabs, or snorts, but it turns into an addiction. We already have gone over the negative impact of addiction, so it's imperative to just say no.

In the university environment, you will be exposed to people indulging in designer drugs and other things such as marijuana to get high, and you're tempted to join their crowd. However, you can have alternate and healthy habits such as sports and other societies. You can also do workouts at the gym, and you will be away from drugs. Hollywood and movies do sometimes unintentionally encourage drugs, so you should never be influenced in any way. You don't need to spend money on these things or get addicted to even painkillers or steroids. Whatever becomes an addiction is bad news. You don't need to do drugs to fit in with the cool crowd because drugs are uncool.

I would go as far as saying that you should avoid smoking altogether, whether it's cigarettes, ganja, weed, or other forms of marijuana. Drug peddlers are always around where you can "score" the "good" stuff from them. They will be in your classes, at the bars, or any public place. Now, here's a catch 22 situation. Cigarettes are easily available, and medicinal marijuana has been made legal for recreational use in certain states in the United States. Therefore, you have easy access to these drugs and can get addicted very easily. Here's the thing: nicotine in cigarettes is an addictive chemical. It may not be nicotine entirely to cause lung problems or cancer, but the tar and the other carcinogens are. Nicotine will, at the least, make you addicted, so it's best to stay away. If you want to quit smoking, there are ways to do so, such as nicotine gums, certain electronic cigarettes, and nicotine-free tobacco and non-tobacco products. None of these is recommended in high doses but could be used as alternatives to quit smoking. I don't encourage using any of these products because they may still have nicotine in them. The nicotine-free products may be a healthier option but should be used for the purpose of quitting rather than starting a new habit. It will give you the sensation of smoking, but at least you won't be impacted by the negative effects of nicotine. The electronic cigarette products at least don't contain the tar and other carcinogens of a cigarette and are reportedly considered to give you five per cent of the damage of smoking as per the Royal College of Physicians[10].

---

[10] The Canadian Vaping Association. (2020, February 6). *Research supports vaping as at least 95% less harmful than smoking.* GlobeNewswire News Room. https://www.globenewswire.com/news-release/2020/02/06/1980703/0/en/Research-supports-vaping-as-at-least-95-less-harmful-than-smoking.html

It would be best to remember that every time you try to quit, don't use these other assisted devices or gums unless absolutely necessary. There are better ways to quit smoking and usually therapies such as Neuro-Linguistic Programming (NLP) to help you quit such habits. Find a company of friends that doesn't smoke cigarettes or marijuana. That's always better because you won't be pressured into smoking and enjoy healthy activities together.

Priorities

If you prioritize things the right way, you can stay away from substance abuse and alcoholism. You have tons of things to do in university, from studies to exams, sports, athletics, fitness, and socializing. If you prioritize your life to give studies and recreational time importance, you'll know that partying and drugs are unnecessary. You will know that these are harmful and will impact your other essential priorities. Fitness, athletics, and sports are all very important because they help you stay fit and strong. They help you develop an active lifestyle and strengthen your immunity.

The best way to prioritize things is to develop a schedule and stick to it. You can speak to counsellors and advisors in university to develop a schedule. They can advise you based on your class schedule and extracurricular activities. This will make sure you're getting at least eight hours of sleep daily and have enough time for sports, athletics, and homework assignments after classes. The better you are at doing this, you will be more used to staying healthy and fit and naturally stay away from substance abuse.

Healthy Diet is a Healthy Life

As a young adult, you're still growing and developing your strength, so you need to have the best diet possible to include all the essential ingredients required to stay fit and healthy. This means boiled eggs, fruits, vegetables, grilled meat, and fewer caffeinated drinks like tea, coffee, and soft drinks. Salads are great, and for drinks, you could go for mint lemonade as it's a healthy alternative to caffeinated soft drinks and excellent as a detox.

Other than that, you can have green tea and decaf coffee too. Black coffee is also good for health. If you stay away from fried food, junk food, and fatty snacks, you're automatically fit. Air frying is better than normal pan frying, but it would be best to stay away from them entirely unless it's part of your cheat meal. The best diet is a sensible diet where you're not required to eat less. You can have decent-sized portions but spread them throughout the day. Avoid food items with cholesterol to stay safe from heart diseases.

The important thing here is to maintain a healthy lifestyle, and food is essential for survival. It builds your immunity and keeps you strong and healthy. Once you develop that mindset, you will automatically stay constantly healthy and away from drugs and alcohol. Your food is your lifeblood, so you need to nourish your body with the right nutrition.

Mental Health

It is said that a healthy mind leads to a healthy body. It's important to stay stress-free at university, considering there's a lot of academic pressure and extracurricular activities students are part of. If you're keeping your mind healthy and stick to a proper schedule, at least you are not suffering from unhealthy stress and anxiety. Stress is known to put people on drugs, and students have to face a lot of it. It's easy to smoke packs upon packs of your favourite cigarette brand because it's so accessible and can offer temporary stress relief. The more your anxiety will be built, the more you will be encouraged to smoke cigarettes or marijuana. If it leads to depression, you could be pulled into even more dangerous drugs. You don't need those in your life at all. There are ways to get around this problem.

When you feel anxious and stressed, you should go to your university's wellness centre and talk to a counsellor instead of taking out your pack of cigarettes and light one up. Try to get an appointment the same day or the next so you can have someone to talk to on how to deal with anxiety and stress. You can even talk to your academic advisors for assistance in helping you build a schedule where you can deal with the academic pressures.

There's no reason to get too pressured because academic stress is normal, and everyone goes through it and deals with it. Counselling, scheduling, and finding time to destress also helps. In your free time, you should get out of your dorm room or hostel, and go for a walk, exercise, or go for a dip in the pool. All of these activities will help you destress and stay relaxed at all times. It's 1000 times better than taking drugs because it will not help you manage academics properly at all.

Parental Pressure

Parents are supposed to be our best friends, confidants, guides, and counsellors. They're traditionally always there for us, but there is a catch. Naturally, they have expectations from us to do well in university as they're concerned for our future. This is why there is a certain level of pressure from them, and we could succumb to it if we're not maintaining a GPA to keep a scholarship or just enough to graduate. If you don't keep them happy by scoring a decent GPA, you're worried that you may not get the funding you need from your parents to stay in university. They may want you to be a valedictorian or get on the dean's list, so you're under even more pressure.

There are ways to overcome that pressure. You can talk to your parents and set realistic expectations and then work towards them step by step. You can move up a notch depending on how you perform each semester. You can take help from university counsellors and academic advisors, and they can guide you on meeting or exceeding those expectations.

If your GPA is tied to a scholarship, you need to mention that to the advisors and counsellors to help you achieve those goals. It's not so hard, and you don't need to fall into the drug trap to get out of it. It won't help and just make things worse. Your parents won't be proud of you if you become a drug addict or substance abuse victim. Those aren't the expectations your parents have from you because that will cause them more hurt and disappointment.

Relationships and Friendships

Getting into university will introduce you to new people, so you will make friends from classes and your roommates. This ties back to the New Horizons section, and we will expand this point further. You will get into relationships with people of the opposite sex, and that does help you stay relaxed. They become your significant other, and you will form strong bonds with them. They will become your companions and best friends, and you will share your secrets with them too. If the relationship gets physical, you must be careful about not getting intoxicated and have unprotected sex. Consent is important, and it's better not to complicate matters if the relationship gets to that point. It's essential to establish healthy boundaries, and then you both will know how to behave as soulmates.

While it's great to have a partner, things worsen when you encounter problems and relationships deteriorate. Drugs and alcohol will come into play as those are easy escapes. Do keep in mind that these will hurt your relationships more than anything else. Never let any relationship get in the way of your normal life and academics because university life is not about having a girlfriend or boyfriend but scoring good grades, learning new disciplines and prepare for the real world.

Strike a balance and have open communication with your partner, so you know your limits and help them overcome your other problems, stresses, and anxieties. This is the best way to keep relationships healthy. The same rule applies to your other platonic relationships and friendships too. There should be normal expectations and boundaries so your relationships and friendships stay healthy.

Be Responsible

I will close out this chapter by talking about responsibility. You are responsible for your actions and decisions, so make wise decisions and consult a good friend, advisor, counsellor, professor, or your parents before making important decisions. You are responsible for your health, too, so make sure you responsibly stay away from drugs and alcohol.

Staying away from substance abuse in university will go a long way in ensuring a healthy life after graduation. To sum things up, we talked about staying safe and protected from substance abuse as adults. Once children reach adulthood, they attain several responsibilities. When they start university life, they are more exposed to situations and people that could propel and push them towards abuse. It's very important to keep oneself protected at all costs at this time. This is a very critical and sensitive time for young adults. They are finding themselves and their career choices. Any slip-up here and they could easily head towards self-destruction. You need to avoid situations where you can fall into the drug trap. In short, stay safe and responsible, and treat yourself and your body well and just say no to drugs and alcohol!

Chapter 10: Reasonable Use and Consumption of Alcohol

Alcoholism is considered to be an abuse of alcohol. Too much alcohol can cause significant mental and physical concerns. Alcohol abuse can be avoided with reasonable consumption. As we move along, we will discuss how the Nigerian Kofyar tribes use fermented beer, which offers health benefits. Alcoholic drinks with relatively reasonable alcohol content such as 5 mg are reportedly not as harmful, and within sound limits, consumption can be controlled hence less chance of abuse. The point is to go from reasonable consumption to abstinence unless it's some social or cultural occasion. If the situation asks for it, then one should drink but within controlled limits and not enough to lose their senses.

The previous chapter focused on how students at university should stay safe from alcoholism and drugs because they are considered young adults having crossed 18 years of age and within three to four years would be entering the real world. Hence, university life is a precursor to the practical life that all adults live, whether they're single or married. When we become adults, we enjoy all the freedom that adulthood gives us, but it presents its set of problems too. Many things then become legal for us, and one of those include alcohol and smoking. In most parts of the world, 18 years and older adults are legally allowed to smoke and drink as they wish. In the United States, the legal drinking age is 21, while it's 18 across the pond. Now, you must have heard and read public service announcements of drinking responsibly. The keyword here is responsibility.

With Great Power Comes Great Responsibility

If you're a Spider-Man fan, you must have heard the saying, "with great power comes great responsibility." When we become adults, we pretty much become free birds and independent souls. We can decide how we want to live and that gives us great power. Ultimately, that power gives us the right to do whatever we want, which means smoke or drink freely. That does require responsibility when it comes to drinking because if you do it too much, you'll face the same problems of intoxication we discussed in chapter 9.

This is where we talk about how we can be responsible while drinking. Drinking is a social pastime, and it's fun too, but you need to know your limits. This means moderate drinking up to the point of not being intoxicated and ensuring you have proper self-control. So, when we ask the question what is considered moderate or reasonable consumption of alcohol, then it means, according to Mayo Clinic, "Moderate alcohol use for healthy adults generally means up to one drink a day for women and up to two drinks a day for men."[11]

Furthermore, according to Mayo Clinic, "Examples of one drink include:

- Beer: 12 fluid ounces (355 milliliters)
- Wine: 5 fluid ounces (148 milliliters)
- Distilled spirits (80 proof): 1.5 fluid ounces (44 milliliters)"

So, when we go above this limit, that would be considered irresponsible consumption of alcoholic beverages.

Return of the Pleasure Principle

We discussed the pleasure principle in earlier chapters which pretty much defines the reason why people get into alcoholism and substance abuse. There is pleasure in drinking alcohol, and the more you get into it, the more satisfaction you derive from it. You want to do more of it so you can enjoy getting intoxication and feeling wild and free. The consequences of that are worse because the pleasure principle is a self-destructive prophecy.

It corresponds with the principle of diminishing returns and diminishing marginal utility. The more you drink, the more

---

[11] *Alcohol use: Weighing risks and benefits*. (2019, October 26). Mayo Clinic. https://www.mayoclinic.org/healthy-lifestyle/nutrition-and-healthy-eating/in-depth/alcohol/art-20044551?reDate=12082021

you intoxicate yourself, and then the satisfaction goes down from there because you start feeling sick. Clearly, the saying "too much of anything is good for you" works here very well. The same goes for smoking. You start with one cigarette to two and then a pack a day. Drinking is a social activity and should be treated the same way instead of using it to lose your senses.

Moderation is Key.

Here we will talk about how it helps to drink in moderation. You'll be surprised to know that moderated drinking actually offers health benefits. One of the key benefits is the reduction in the risk of developing heart disease and even dying of it. This itself was quite surprising as drinking too much reportedly leads to liver damage. Furthermore, you can possibly reduce the risk of ischemic stroke. According to Mayo Clinic, that happens when arteries to your brain get narrowed or blocked, causing severely reduced blood flows. You can even benefit by reducing the risk of diabetes. These are excellent benefits of moderated drinking, so that's an incentive for anyone who enjoys drinking beer, whisky, wine, or any alcoholic beverage.

At the same time, it's more important to consider less risky indulgences and activities such as a healthy diet and being physically active. These are two more proven and successful ways to keep fit and stay healthy, plus it gives you an active lifestyle. All of this is way better than moderated drinking because it's hard to stick to "moderation" and not going overboard. The temptation is always there, and the benefits can turn into problems.

Heavy Metal

The risks outweigh the benefits despite the benefits being quite tempting enough to get people to drink moderately. Things get messy when moderated drinking turns into heavy drinking. While benefits of moderated drinking are noted, there are absolutely no benefits in heavy alcohol whatsoever. It is, in fact, a death sentence because you're doing so much damage to yourself without knowing. When we define heavy drinking, it means having over three drinks a day or over seven drinks a week for women and also for men over 65 years of age. Furthermore, heavy drinking applies to having over 14 drinks a week for men 65 years and younger, as stated by Mayo Clinic.

If you think heavy drinking is the maximum limit for drinking, think again. There's even binge drinking that is the norm in university parties. Mayo Clinic defines it as having four or more drinks in the span of two hours for women and five or more within two hours for men. Binge drinking essentially means excessive drinking, and that's when you will lose all sense of control and will become completely feral. This is dangerous because the risks are not worth the pleasure at all. The pleasure principle exists here, but too much is way too much.

When you indulge in binge drinking, you incur serious health risks that include specific cancers such as breast mouth, throat, oesophagus, and liver cancer. You can contract pancreatitis too. Furthermore, you can face sudden death, especially if you suffer from cardiovascular disease. You could also suffer from heart muscle damage which is known as alcoholic cardiomyopathy. That eventually leads to heart failure. So, if you look at the risks so far, they're mostly related to heart and cardiovascular problems. It doesn't stop here as other consequences of binge drinking include stroke, high blood pressure, liver disease, suicide, accidental severe injury or death, brain damage, and other problems in an unborn child. Finally, you also risk alcohol withdrawal syndrome.

When to Stay Away

Of all the risks mentioned, it's hard to place one over the other as they're equally life-threatening. You could cause serious injury to yourself and others because you're too intoxicated to drive or be left to your own devices. Excessive drinking can be a result of depression, and that could potentially lead to suicide too. Brain damage would make you pretty much a vegetable because we can't do any of our regular tasks without a working brain. The brain is our body's central processing unit, aka CPU, so damage to that will trickle down to your entire body's functions.

We have discussed the benefits of moderated drinking and the risks of heavy and binge drinking. Keeping all of these in mind, you should know when to avoid drinking. There are situations in your life that totally rule out even moderated drinking. You should confirm with your doctor if you decide to indulge in drinking at any time, but it's most important to avoid when you're pregnant or want to have a baby. Suppose you are already addicted to alcohol or diagnosed as an alcoholic or run in your family. In that case, it's best to avoid it altogether because you could fall into the same trap. If you have had a hemorrhagic stroke, then you must stay away from drinking. This is caused by leakage or rupture of a blood vessel in your brain. Remember how we mentioned that brain damage is a risk of binge and heavy drinking, so it's only advisable to abstain. If you suffer from cardiovascular and heart problems, then drinking will only make things worse. If you have had heart failure previously, then drinking will only make you vulnerable to it again. Furthermore, if you are prescribed over-the-counter medications that can interact with alcohol, it's better to stay away from drinking.

In all honesty and sincerity, I would advise against drinking altogether because the risks outweigh the benefits. Moderation can potentially lead to heavy and binge drinking, and the risks will be way too much to handle. If you really do feel the urge at times, it's best to indulge only in moderated drinking and be responsible because your health is your responsibility. As always, it's best to consult your doctor before you plan to start drinking. Your health and safety come way before the pleasure principle.

Nigerian Revolution and the Kofyar Beer

According to research, Nigeria presents an excellent model for positive drinking. In fact, it's become so in tune with the culture of the Kofyar tribes that drinking is not given much importance at all. There's little or no drinking for the most part because the emphasis on addiction is very less.

The Kofyar tribe hails from Northern Nigeria and consists of farmers lacking occupational specialization. Their lives revolve around beer because they are constantly making it or drinking it and then having thoughtful conversations about it. It's a lifestyle for them, and they take pride in presenting beer as a mark of esteem and affection. It's so much ingrained in them that friends and lovers are always seen together drinking simultaneously from the same calabash. Celebrations involve harvest beer made by individuals or popularly subscribed and then freely distributed to others. Beer is presented as a token for socially valued roles. Whenever a man gets married or pays his taxes, he is then entitled to take extra beer with him home.

Research states the following about the Kofyar tribe:

> "The Kofyar value mild euphoria, the sense of happy relation which accompanies drinking. Yet, there appears to be no particular premium placed on increasing either the proportion of alcohol in the brew or the speed of its consumption. The alcohol content of the beer averages three to five percent. This is rather less potent. The Kofyar

> people in northern Nigeria could probably get more alcohol into their bloodstreams if they increased the speed of drinking. The actual drinking may be quite rapid, but there are often waits of an hour or more between drinks. Also, the Kofyar people value their local beer as a social lubricant but not as an explosive. Drinking is understood to be a social act."

For the Kofyar tribe, it's not about consuming alcohol but enjoying beer and using it for festive purposes. Its purpose is more to enjoy the taste and present it as gifts or for ceremonies. They even have beer parties, as explained below:

> "The beer party among the Kofyar also acts as a proper occasion for the discharge of interpersonal tensions for public review. A man who customarily shouts and roars on his way home from a beer party is regarded as ludicrous rather than disgraceful. In Horton's terms (1943:266), the Kofyar are characterized by slight insobriety in which alcohol is used regularly but with restraint. Stupor is neither the desired goal nor the usual outcome of Kofyar drinking."

Even though alcohol is a primary component of beer, the Kofyar tribe knows how to drink in moderation. Whatever we've discussed on the benefits of moderated drinking, they're already benefiting from it because they are not overdoing it. It's a part of their culture and lifestyle, so they don't seem to find pleasure in getting drunk. There are rules in place because they consider alcoholism a learned behaviour in their culture and environment where the alcoholic was raised in. There is absolutely no room for violent behaviours following drinking. Therefore, heavy and binge drinking chances are low because the tribespeople observes the local drinking laws. As stated below,

> "The culture dictates that disputes must be resolved in a public arena for all to hear and review. Whereas in other cultures, one is allowed to drink to forget his problems which in turn leads to uncontrollable drinking to suppress inner tension. Loss of control could be argued as a way to avoid dealing with tension. Also, loss of control drinking could be argued as due to addictive chemicals in industrial alcoholic beverages for the alcoholic to drink more to increase sales. Also, the high percentage of alcohol in alcoholic beverages is too strong for the body to process and metabolizes, i.g. wine, rum, etc., made out of fruits."

Since we've discussed so much about the Kofyar beer, you must be thinking about what's in it that makes it so potent. It's not really distilled liquor but fermented local grain that contains small quantities of alcohol. Therefore, there is zero loss of control drinking among their people. This is because they follow strict rules regarding abstaining while drinking. They reportedly abstain for at least an hour before indulging in their next drink during a party or festive occasion. Their parties aren't university or club parties but more ceremonial events. Basically, distilled liquor, in contrast, has a higher concentration of alcohol that is considered way too strong for the human body. Hence, the Kofyar people don't consume hard liquor as it won't serve their purpose as a social catalyst, ceremonial events, festive occasions, and celebrations.

Drink Responsibly

As the example of the Kofyar tribe shows us, we can enjoy alcoholic beverages in moderation or those that are fermented and low in alcoholic content. We should eliminate the need to get drunk or wasted and enjoy these beverages for their taste. It's vital to exercise caution and responsibility because the pleasure principle will persist here too. Practising abstaining is essential like the Kofyar people do, so it's imperative that you follow their example. Drink responsibly for the safety of yourself and others around you so you don't fall into the trap of alcoholism.

Chapter 11: The Real World

It is imperative for adults that they protect themselves from substance abuse after entering the real world on graduating from university. They may start a business, get employed in a company, or get married and start a family. In all situations, they will have responsibilities and deal with financial pressures and stresses. All these situations could turn themselves towards drug addiction and eventually substance abuse if not handled well. This is why this chapter will focus on staying safe and protected while balancing our work and family lives as adults.

It has been said quite a few times when students graduate from college by their parents and professors, "Welcome to the real world!" It is very true that when we graduate from college and university, we pretty much enter what is known as the real world. How do you define the real world? The real world is called practical life, adulthood, or the time in a person's life where they assume responsibility for their finances, professional and personal needs, and goals. There isn't much to it because it's a rite of passage all of us go through. Being an adult is a privilege because it gives us responsibilities but also allows us to live our dreams. We can have adult relationships, get married, have a family, travel, and see the world. We can have jobs, earn a living, and enjoy fulfilling lives. Once again, I'd like to reiterate that with great power comes great responsibility. This is something I have emphasized several times in this book. This is because we all have a responsibility to ourselves and our future generations to be good human beings. We need to set an example for our children and their children to follow. With that being said, we need to be very vigilant with ourselves, so we can stay away from going on the wrong path in our lives. This is why we are independent decision-makers as adults. We aren't answerable to anyone in our personal lives but only to our seniors in professional lives. Other than that, we have to be careful not to indulge in self-destructive activities. One of those is substance abuse, so it's only beneficial to us that we stay on the right course and protect ourselves and our family from its curse and stay strong against all such temptations.

Stay Safe

The first rule of thumb is to stay safe. Staying safe means keeping yourself away from all drugs, substances, and alcohol, so you never indulge in them. Alcoholic beverages are fine in moderation, but overindulgence is never good, and we have discussed this in quite detail in the previous chapter. We have gone through all the health benefits and costs, so you know it's important to drink responsibly. Your health comes first, so do the right thing.

Staying safe would mean keeping away from any social company that drives you towards drugs and alcoholism. The right social circle indulges in healthy habits and does not overdose on drugs or partying too hard. This is why your social circle from your workplace and otherwise need to be people who don't abuse drugs or alcohol. If you are part of it and succumb to peer pressure, then the game is over. You should know that drugs aren't cool, and you shouldn't use them as social connectors or find a group of friends.

Your life will no longer be the same, and the impact on your family will be even greater. We have gone through the impact on family and society in previous chapters, so you can see how dangerous this can be. You're not going to help matters because your professional and personal life will crumble around you. The best way is to stay safe so you can have a normal life. No matter what the circumstances are, stay away from substance abuse at all times. Stay safe, so your family and your professional life are protected. If you can't help drug addicts and substance abuse victims stay off their substances, then don't join their cause because you'll hurt yourself the same way they're hurting themselves!

Seek Help

There's the thing when it comes to our lives these days. There's a plethora of stress we all have to deal with, and it drives us insane to the point that we can get exhausted and burned out. Burnout is not the best outcome, and it can lead to disaster such as a mental breakdown. When you get to that point, you will want to indulge in drugs and alcohol to get rid of the stress. This is because you will become so burned out and stressed that you will see no other way to deal with the stress. When you come across such situations, you don't need to panic but just seek help. Any kind of therapy, counselling and assistance from friends and family will help the situation. It's the only way to go because drugs and alcohol are only going to push you to the path of self-destruction. There is no worse thing to do than to hurt yourself even more. These drugs aren't going to help your cause at all. You will plunge deeper into depression, and you'll find no way to get out of it if you reach a state of complete helplessness.

You need to understand that you have to survive for your friends and family. Seek help from the right sources and not drugs, and you'll fight your way out of your depression. Everyone faces challenges in their lives, and we deal with them differently. When we face these situations, we need to know that this too shall pass. We should be brave and strong to face them instead of being weak and coward under pressure. If we take the right help, we will know how to tackle these challenges bravely. The right counsellor will guide us and point us solutions to our problems. That doesn't need to be professionals but friends and family too. There is support around us, and we don't need to go down any dark path. As discussed earlier, we will only hurt our families, and we have seen how worse the damages are to our family. Please don't harm yourself anymore but seek the right help. This isn't just a request but a plea for your sanity.

Work-Life Balance and Active Lifestyle

When we reach adulthood, our lives become so fast that we're on a runaway train, and sometimes we just can't catch up. Our work stresses transfer to our family, and that's why we need a work-life balance that is very balanced. Balance is the key to ensuring a healthy professional and personal life. By not exerting pressure on yourself via work, then you can give equal time at home too because you need your "me" time and family time. You need to give yourself time by adopting a healthy lifestyle. You can do that by having a regular schedule of work hours and time for exercises and fitness, and then family time.

When you work on a fitness routine, you let go of all your stresses and focus your energies on rebuilding your body. We let go of all our stresses and tensions because exercise acts as a tension release, and it's imperative to have that healthy and active lifestyle. It also means eating right because that gives you the right energy to be productive at work. You need to build muscles and energy to become a stronger person inside out. By doing this, you will only do yourself a favor. If you want to stay off drugs and alcohol, you need to practice restraint and focus on yourself. Once you focus on yourself, you'll realize you don't need the help of drugs and alcohol to survive.

Find a gym nearby and get a membership. Take group classes and help from a personal trainer because both will help you work on yourself and get in touch with fitness-minded people just like you. That's the circle you need to be in because, as we stated before, the right company is the best way to stay safe. When you build a schedule and stick to it, you'll work harder to not deviate from it and therefore avoid the risk of exhausting yourself and burnout. Make sure to include the three key things we discuss, such as eight hours for work, an hour for fitness, and the rest for the family. If you sleep on time and have eight hours of sleep, you'll be more energetic every day.

Dependence

One thing I must stress on here is that never do too much of one thing that it becomes an addiction. You don't need to get dependent on any drugs, medicines, cigarettes, or alcoholic drinks. This requires a change in mindset, and you need substitutes. As we stated earlier, get dependent on fitness routines and develop other healthy habits. You could go trekking, hiking, cycling, and do light jogging. Power walks are very helpful. Find hobbies that challenge your creativity. Spend time playing mind games such as chess or do puzzles in newspapers. Adult coloring books are great stress-relievers too. There are a lot of things you can do to keep yourself busy and not dependent on any substance.

Build a group of friends that help you inculcate such habits, too, because these are extremely helpful. When you see more like-minded people doing healthy activities, the more you'll be encouraged to partake in it. You can plan road trips and vacations too. You could be part of book clubs or pursue creative passions such as writing, blogging, video blogging, or study new courses in science, technology, business, and creative fields to keep yourself occupied. You do realize that an idle mind is the devil's workshop, so make sure you're not wasting a single minute of your day doing things that are unhealthy for you. A healthy mind leads to a healthy body, and the drugs impact your mind first and make them dependent on them. For example, nicotine is an addictive chemical found in tobacco products such as cigarettes and cigars. There is also a huge connection between dopamine and nicotine. According to research, "Nicotine that gets into your body through cigarettes activates structures normally present in your brain called receptors. When these receptors are activated, they release a brain chemical called dopamine, which makes you feel good. This pleasure response to dopamine is a big part of the nicotine addiction process."[12]

Remember how we always go back to the pleasure principle? This is exactly what that entails because nicotine helps release dopamine, and that does make you feel excited and happy. The more you smoke cigarettes, the more you'll get addicted, and the chances of you contracting lung cancer and other dangerous diseases are even higher.

---

[12] *Smokers' Brains Change in Response to High Levels of Nicotine.* (n.d.). Https://Newsnetwork.Mayoclinic.Org/. Retrieved August 21, 2021, from https://newsnetwork.mayoclinic.org/discussion/smokers-brains-change-in-response-to-high-levels-of-nicotine/

Please take a look at the following research and see why the dopamine-nicotine connection is so self-destructive:

"Over time, as you continue to smoke, the number of nicotine receptors in your brain increases. Addicted smokers have billions more of these receptors than nonsmokers do. But not all smokers have such a high level of receptors. That is why some regular smokers can stop smoking without much difficulty.

When you try to stop smoking, the receptors in your brain do not receive nicotine, so the pleasure response is cut off. In addition, low levels of nicotine lead to symptoms of nicotine withdrawal, such as strong cravings for a cigarette, anxiety, irritability, restlessness, difficulty concentrating, depressed mood, frustration, anger, increased hunger and difficulty sleeping. The fastest way to alleviate the withdrawal symptoms is to smoke a cigarette, which releases dopamine and activates the pleasure response.

To make stopping smoking even more difficult, the brain receptors can be conditioned to expect nicotine in certain situations long after you have stopped smoking. For example, if you regularly smoke when you drink alcohol, or when you are in a stressful situation, or after a meal, the nicotine receptors in your brain anticipate the dopamine rush from nicotine at that time.

These "trigger" situations can cause intense cravings for a cigarette, even if you have stopped smoking for several months."[13]

This is how cigarette commercials back in the day made you feel cool such as the concept of the Marlboro Man. Drugs aren't cool, and cigarettes are easily available, and they're only one of the worst health hazards you can adopt as an addiction. It's so bad it's good, and that's never an amazing thing. You can release dopamine naturally by adopting healthy habits. If you're a fitness freak, you'll get dopamine releases by being pumped up and confident. You'll even feel accomplished because you achieved certain fitness goals. This brings us to our next section, and that's goal setting.

Goal Setting

We've spoken about fitness goals, activities, and healthy habits, but goal-setting adds another layer to this. The way I see it, it's not about setting goals every week or month but every day. You can plant those in your daily schedule, so you have something to look forward to. These could be self-contained goals like reading a novel in two hours or writing a blog. These could also be micro-goals leading to major goals such as a fitness routine to lose weight or build muscle mass.

---

[13] *Smokers' Brains Change in Response to High Levels of Nicotine.* (n.d.). Https://Newsnetwork.Mayoclinic.Org/. Retrieved August 21, 2021, from https://newsnetwork.mayoclinic.org/discussion/smokers-brains-change-in-response-to-high-levels-of-nicotine/

Other goals would include working towards a promotion at work or salary increments. This way, you'll be more focused at work and set milestones to grow professionally. Both personal and professional growth are equally important, so you need to note down goals in both areas and work on them.

Make goals manageable by breaking them down so you don't stress yourself out too much. These could be as simple as buying a new house, a car, or going on a vacation with loved ones or spouses. It could be proposing to a significant other and planning a wedding. It could be celebrating milestones or past achievements. Plan it out, set start and end dates, and work towards them no matter what your goal is. If you want to quit smoking, that's an even great goal, or moderating drinking or going completely clean and sober. Either way, these goals are productive and what we would term as SMART, which stands for Specific, Measurable, Achievable, Realistic and anchored within a Time Frame.

So, for example, you want to quit smoking, so that's being very specific. You can measure this by going from five packs a day to three and so on. It's very much achievable to quit smoking and, at the same time, pretty realistic. Finally, you can plan to quit smoking cold turkey in a span of three months to a year or however you see fit. There's no rocket science to it, so think SMART and set SMART goals!

Support Groups

We talked about healthy social circles, but sometimes you can take the help of support groups where you can meet people facing similar issues or talk about different issues every week that you all can tackle together. These could be custom groups focusing on depression, anxiety, stress, or a combination of three, for example. These could also be formal groups such as Alcoholics Anonymous. You can form a support group of friends who want to quit smoking and control their drinking habits, for example. There are several ways to go with support groups, but you must find a group you can identify and fit well in and then find solutions to your problems. There are groups for divorce, relationships, breakups, and more. You name it, and you got it because you can simply go on Facebook and find groups to meet people and set up physical meetups. It is so easy now to have support groups than it was before. You just need to know where to look. The resources are there, and it's your responsibility to use them!

Welcome to the Real World

"Welcome to the real world. There's so much to learn..."

The above is from a song by 80s-90s pop-rock outfit Mr. Mister, and it's no further from the truth. The real world offers you several learning opportunities. You can make mistakes but learn from those experiences, good or bad. You will learn from all the challenges you've faced. The real world is tough, but I am confident you'll survive without being a substance abuse victim. As part of my parting thoughts on this chapter, I will leave you with some words of wisdom from the Italian Stallion Rocky Balboa, as portrayed by Sylvester Stallone in the Rocky movies. This was from the sixth movie in the franchise titled "Rocky Balboa," where the former boxing champion tells his son how he should not blame him for living in "the father's shadow" and be a man! You should memorize them because these golden words are the key to beating life and its challenges and being a winner at whatever you do!

> "Let me tell you something you already know. The world ain't all sunshine and rainbows. It's a very mean and nasty place, and I don't care how tough you are it will beat you to your knees and keep you there permanently if you let it. You, me, or nobody is gonna hit as hard as life. But it ain't about how hard ya hit. It's about how hard you can get hit and keep moving forward. How much you can take and keep moving forward. That's how winning is done!"

Chapter 12: Treatment Plans

In some cultures, alcohol and substance abuse is looked down upon, and those who are part of it may not want to seek professional help. Society considers it taboo, and families and their children may not be inclined to speak to therapists or counsellors. These people dealing with substance abuse are afraid that they will be judged, so they don't choose any remedy or counselling to get over it. This is why raising them right is the way to go. Parents need to be educated on keeping their children safe from substance abuse, so they don't go down the path of self-destruction. Mental health is a stigma, and not everyone takes it seriously. We need to make tremendous efforts and strides to make our families, friends and loved ones aware of mental health issues and how they should be taken seriously. We should help those who are struggling with substance abuse or have preventive measures in place so that they never go down that road. Therapists need to understand the culture of those dealing with both alcohol and substance abuse and adapt their counselling to deal with the cultural aspects.

Whenever you're sick, you seek a doctor for treatment or take medicine for the pain or illness. When it's a physical illness to worry about, you will have a remedy for it most of the time. For example, you can take any anti-allergy or antibiotic over the counter to treat the common cold. Tylenol is almost always used in America to treat the common cold, flu, or anything else. Painkillers are recommended for various illnesses. Now, let's flip the switch here and replace physical with mental illness. Mental health is a critical concern in today's fast-paced world. It's no secret that stress is one of the significant causes of heart disease. Stress causes anxiety, and that impacts the heart. The body and mind are connected like yin and yang. These two elements are so interlinked that you can't have one without the other. I've stressed many times in this book how a healthy mind leads to a healthy body.

Now, the reason why I am mentioning mental health here is because of its connection to substance abuse. The drugs have certain chemicals in them that give some sort of short-lived pleasure, but they are highly addictive. These release a lot of dopamine, and that's when you feel that high or buzz. Take, for example, the chemical nicotine that is found in several tobacco products, from cigarettes to cigars to pipes to electronic cigarettes, otherwise known popular as vapes. The chemical releases dopamine, and that's when you want to keep doing it more and more. Cigarettes, as already noted, are easily accessible and available as long as you are of legal age. Some young minors get into it by hanging out with adults and getting caught in the habit. When they get into the habit, it starts a vicious cycle, and that creates a lifetime of addiction. It's not easy to kick the habit, and when you've kicked it, it's easy to fall back in that trap too. Withdrawal symptoms never make it easy to stop smoking and even drinking, for that matter. Cigarettes are the easiest and perhaps the worst form of addiction, but the health risks are very high. Alcoholism is similar because it's also easily accessible and available if you're of legal age. Drinking alcoholic beverages makes you drunk, which in other words, is losing your senses and self-control, among other things. When you lose grip on yourself, your mental health is sadly impacted. This is why alcoholism is one of the worst forms of substance abuse.

Whether it's alcoholism or cigarette smoking, it's substance abuse. When anything has abuse attached to it, it becomes a deadly disease because it impacts your health enough to die. If you suffer from a sickness, you need to be treated. Hence, there should be treatment strategies and plans to prevent this disease because simple medicine isn't the answer to these problems. Effective treatment plans are available, and those should be looked into when you or anyone you know wants to kick the habit.

In chapter five, we discussed various treatment strategies that included several therapies that we had discussed. The treatment plans discussed in this chapter will tie into what we had talked about earlier. For example, if you plan to quit smoking, you can speak with a therapist and develop an effective plan to quit smoking within a specific time. For example, you may have to take up cutting down on smoking from three packs a day to one and so on. There is no one single way to achieve kicking the habit, but it is possible.

You will have to try different plans that work for you, including reducing the habit and participating in various therapies. NLP is one strategy that has been reportedly useful in helping quit smoking. It goes without saying that substance abuse treatment is a lifetime plan because it impacts your entire daily life, including your personal and professional progress. There are various steps involved in the treatment plans process, and we will outline those below.

Diagnosis

Before any treatment plan can be prescribed to a substance abuse victim, the health care provider has to conduct a diagnosis that involves several things. The diagnostic review includes substance abuse patterns, the patient's medical history, and any recurring, past, or current mental health conditions. A medical checkup can be done to see how healthy the patient is and whether they would require medications or not. Once the diagnostic review is conducted, the medical health practitioner can come up with a summary of the major problems that motivated the patient to take up treatment and what treatment plans and recommendations they should follow, including a mix of medication and behavioral therapy.

Without isolating the issues and looking at medical records and patient histories, it is difficult to recommend a treatment plan. Every patient works differently, so the treatment plan has to be customized for it to be highly effective in treating the patient's substance abuse dilemma.

Listing down and zeroing in on the problem

Anyone suffering from substance abuse has several problems that they face related to their habit. It could be smoking over three packs a day or getting drunk daily. It could also be the inability to reduce alcoholic beverages intake or spending too much money on cigarettes in a week. The problems need to be listed by their symptoms, such as arrest while driving recklessly or under the influence or being arrested for drinking while driving. It could also be recognized by analyzing drinking habits, for example. This is because the mental health provider or doctor, or therapist can realize the problems resulting from substance abuse and explain the causes and evidence. This way, it can work on those specific problems and habits and hope to reduce the habit and subsequent abuse.

Goals

We had earlier discussed goals in the previous chapter and emphasized SMART goals that meant that they had to be specific, measurable, achievable, relevant and time-bound. A quick recap on those goals means that whatever plan you have in mind or the patient does, in terms of quitting drinking or smoking, has to be specific, measurable, achievable, relevant, and time-bound. Any kind of addiction terminating can be achieved via these SMART goals, and that's an assessment the doctor or therapist can make with the patient and make a plan on how to achieve them, keeping the above parameters in mind.

Every problem has a solution, so you need to be SMART when it comes to finding those solutions. Goals need to be set, and they could be just a statement as simple as "I want to quit smoking" or "I don't want to be an alcoholic." These are all SMART and can be achieved within a time frame. Your goal has to be related to the substance abuse problem you have. You have to confirm within your mind that you have to do it with the guidance of your therapist or doctor.

Goals can be broad, too, because they will allow you to replace any unhealthy habit with a healthy one. Substituting addictions with healthy habits such as exercise or some other recreational activity is very helpful. You need to work on yourself and keep a time frame in mind. This is why SMART goals work best in these endeavours. Your goals could be as simple as even overcoming anger and managing your mood swings because drugs cause that a lot. Either way, work on each plan one at a time. NLP is usually beneficial in goalsetting, so you can look into it. While you're setting goals, make sure you have an agenda in mind to achieve your goals. It could be sticking to a diet programme, or Alcoholics Anonymous meetings, therapist appointments, acting on their guidance and following through on their recommendations. It could also be not missing any medications as well.

Healthy Interventions

Interventions are healthy if they're used to help you fix your problems and work on yourself. These could be tough love from family and friends because addictions can cause a lot of self-harm. Your treatment specialist, therapist, or doctor can be beneficial in that too. First of all, you need to identify the problem. It could be related to drinking problems or emotional health. It could be any mental health-related issue as well. Once you place the problem, you have to have a goal in mind, and it could be as simple as managing stress or cut down on drinking or smoking. Then you need to identify the objective or way of achieving that goal. Those could be attending Alcoholics Anonymous or other support group meetings because it helps to relate your problems with others and work on problem-solving together.

The final aspect of the treatment would be to be part of an intervention meeting. Your treatment specialist, doctor, or therapist will refer you to different support groups, help keep a check on your weekly meetings and such, and then assist you in fixing issues in attending those weekly sessions or whatever hurdles you face. These could be anything from transportation, logistics, or the provision of daycare and childcare services, so your children are taken care of. It could be working with family and friends to help attend those sessions too.

Work in Progress

Evaluation and progress checks are part and parcel of any addiction treatment plan. When we talked about SMART goals, we talked about how they need to be measured. Progress needs to be measured and evaluated at all times. Your doctor, therapist, or treatment specialist will constantly be evaluating progress and see if the prescribed treatment is working or not. Otherwise, the specialist would need to advise another treatment or support plan. It will include responses to treatment and seeing if the addiction problem is reducing or not.

You would be asked to write your feelings and thoughts in a daily diary. Putting down thoughts and feelings is a great way to let go of your emotional traumas and not keep them bottled up inside. You can write about why you enjoy the addictions you have and whether you want to get rid of them or not. It's a great stress reliever when you write your thoughts and feelings down. By keeping everything bottled up, your stresses and anxieties would only get worse.

Write your thoughts and feelings down so you can reflect on them later when you get free of addictions. It's unbelievable when you look at the results. You'll thank yourself and your specialist later. Whether it's a mental health practitioner, therapist, or any doctor, he needs to treat you right, so you need to evaluate him as well. If any specialist isn't working for you, then you have the right to change them. It is imperative to work with a specialist who genuinely values you and sincerely wants to change your life. Choose wisely, and you won't regret it. Find a middle ground, and you'll embark on the right path.

Effective Long-term Care

As mentioned above, any drug addiction or substance abuse will require effective long-term care. When working on treatment plans with your physician, specialist, doctor, therapist, or mental health practitioner, you need to work out an effective plan that adequately deals with your addiction and abuse problems. One more thing to keep in mind with any addiction or substance abuse problem is the chance of a relapse. Whenever you've overcome any addiction, withdrawal symptoms are very normal. You need to cater to them effectively as they could pull you towards addiction once again.

Your plan needs to be continued until you're entirely free of it and there's no chance of withdrawal symptoms. It would be best if you discussed long-term treatment with your health care provider, whichever form they may take. Continued long-term care would include attending group therapy sessions for as long as it takes to be rid of all withdrawal symptoms. You need to partake in individual counselling and therapy sessions because group therapy is just one facet of it. Furthermore, it would help if you continue taking prescription medications or any medication-assisted treatment that is required.

Nothing is Impossible.

Substance abuse is a disease. Like any disease, it has a cure. The cure isn't simple, but it's effective and requires long-term treatment plans, so it's up to you how you take action. You need to find a specialist to work with, map out SMART goals and strategies to achieve them. Set measurable targets and agendas, attend support group meetings, counselling, or whatever else it takes to get rid of the problem. That may or may not require medications, and that also depends on the severity of the problem. As with everything else in life, nothing is impossible, and you can kick the habit if you want to. This applies to everyone reading; whether it's you or your loved one facing a problem, don't let go of your hopes. You can tackle it, and you will come out a winner!

Made in the USA
Monee, IL
03 November 2026

67376115R00085